Tropical Tasar Culture in India

Prafulla Kumar Mohanty

2013
Daya Publishing House
A Division of
Astral International Pvt. Ltd.
New Delhi – 110 002

First Published: 1998

Reprint: 2013

ISBN 978-81-7035-189-4

Published by : **Daya Publishing House**
A Division of
Astral International Pvt. Ltd.
81, Darya Ganj, Near Hindi Park,
Delhi Medical Association Road,
New Delhi - 110 002
Phone: 011-4354 9197, 2324 8134
Fax: +91-11-2324 3060
E-mail: info@astralint.com
Website: www.astralint.com

Laser Typesetting : **Classic Computer Services**
Delhi - 110 035

Printed at : **Salasar Imaging Systems**
Delhi - 110 035

PRINTED IN INDIA

Acknowledgements

I wish to express my sincere gratitude to my research guide and supervisor, Dr. Milan Kumar Behera, Head, P.G. Department of Zoology, G.M. College (Autonomous) Sambalpur, Orissa for rendering constant guidance and supervision throughout this course of study.

I deeply appreciate Prof. M.C. Dash, Ex-Head of the Department, Sambalpur University, Jyoti Vihar, Burla, Sambalpur, Orissa for his help and valuable suggestions rendered to me.

I am also very much indebted to Mr. P.S.S. Thomas, I.A.S., Member Secretary, Central Silk Board, Bangalore ; Director, Central Tasar Research and Training Institute (CTR & TI), Ranchi; Mr. B.K. Chakravorty, Joint Secretary (Technical), Central Silk Board, Bhubaneswar, Orissa ; Director of Textiles, Bhubaneswar; Mr. B.K. Nayak, Asst. Project Officer, Bivoltine Sericulture Development Project (B S D P), Chandragiri, Paralakhemundi, Orissa and Dr. B.K. Dikshit, Senior Research Assistant (SRA), Basic Seed Multiplication and Training Centre (B S M T C), Baripada, Orissa for the help they have rendered in providing field information and basic literature which has facilitated the completion of the investigation in time.

My thanks are due to the National Centre for Science Information (N C S I), Indian Institute of Science, Bangalore for providing Current Awareness Service from time to time regularly, and M/s DATA INFOSYS, Bhubaneswar for timely computer service in preparing the manuscript.

Lastly, I owe my thanks to my twin daughters Juli and Mili and my wife Littli but for whom it would have never been possible to wind up the work.

Prafulla Kumar Mohanty

Preface

Today man has made a herculean progress in various fields of science and technology. This has been possible because of his serious exploitation of the nature and natural resources. Of course, this exploitation is attempted exclusively for pleasure and comforts which has resulted a global disaster. Human beings and scientists in particular have forgotten that they are a particle of the nature, born and brought up in the natural environment, which should not be denatured. But the people in and around the nature have dissected this natural environment in such a fashion that the nature has faced the ecocidal situation.

Fundamental biological research in the new century will probably have a radically different perspective than what it was even a decade back. Commercialization of knowledge arising out of basic research has now become the driving motive for many of the scientific activities in most parts of the developed world. Scientists are certainly engaged in magnanimous research, but the result is mostly confined to the laboratory itself instead of releasing to the field and common people. The objective of true scientific research is that it should cater to the need of common people. Does it happen so? Furthermore, biologists are now engaged in research in which the nature, natural system and processes are greatly manipulated by various permutations and combinations. The explosive strides of high yielding variety of rice, hybrid fruits, vegetables, animals etc. are all in front of us which are the contributions of our bioscientists. But nobody probably knows the result after using all these substances. In fact, time will prove its effect and nature.

Sericulture has attracted special attention of India since Independence due to rising internal demand and its export potential. Sericulture and manufacture of silk products have long significant development in terms of productivity and quality right from cocoon production to processing of silk due to sustained efforts made by different agencies. Sericulture is the combination of people, machine, instruments, equipments, raw materials, proper methods and environment. Over the decades, the production of silk in India has increased horizontally, but its vertical growth remained retarded. Against the increased demand for quality silk in the international market, scope for production of export quality silk in India is enormous.

Although there is a considerable literature on sericulture, yet information on tasar culture is scattered. However, the lack of adequate research and text book required by Post-Graduate and research level of Indian Universities stimulated the author to prepare this book. In this book emphasis has been given on the techniques required for the better quality and quantity production of tasar cocoons through systematic and natural process instead of adopting pure scientific equipments or manipulating the natural system of operation. Hope, this process would be ventilated to the farmers for the successful output by effective management. This would not only be helpful to all segments of people of the state or country but also accelerate the country's economy.

I shall consider my efforts to be amply rewarded if the book serves the needs of the researchers, students and farmers in particular to the best of their advantage. Suggestions for further improvement will be thankfully received and acknowledged.

Prafulla K. Mohanty

Contents

1
Introduction

1.1. Importance of the Study

Silk, as a fibre, spells the class of excellence due to its sleek nature, brilliant texture and natural lustre. The touch of silk on the fingertip and its garment on the body really evokes the very thread of its history, a shimmering fabric of far places. Mankind has always loved silk, which has always an aura of royalty. Silk being the most exquisite textile fibre, continues to enjoy a unique place and has always challenged any other textile because of its natural glamour. In fact, it is the yarn of life, extruded by that unassuming caterpillar in a continuous fine filament as long as a mile. Anybody using any piece of silk garment cannot simply believe that the product is just the summation of some fibres produced by tiny worms in nature. It cannot be expressed in few words, how and why is this precious fibre secreted from the mouth of a creature in an intertwined manner.

Silk is named in legend, fable and folklore. The word spells luxury and class. The superiority of silk as a textile fibre has been recognised from time immemorial and even today no other fabric can match its lustre and elegance. Mankind has always loved silk. It has been popular for its unique appearance which can be described only in terms of its aesthetic appeal. This appeal is due to its peculiar lustre, feel, suppleness, crispness combined with softness, resilience and various other properties possessed by silk fabrics. Large number of man-made fibres have since captured the market which was earlier monopolised by three main natural fibres : cotton, wool and silk. The initial attempt was aimed at obtaining *artificial silk,* in other words a fibre resembling silk. Today after having a variety of man-made fibres, it has become clear that there is not a single fibre which can claim to be exactly like silk or which can even be very much like silk. Therefore, silk has proved its own uniqueness of its kind in textile industry.

The advantage of man-made fibres like comparatively low cost, durability and wash and wear properties have not outweighed the indisputable superiority of silk regarding softness of feel, fineness, smoothness and above all its incomparable affinity for dyes which are considered very important in textile industry. There is no other fabric that can really match the beauty of silk. The luxurious look, smug feel and the loomy lustre of silk is unquestionably inimitable. It is regarded as the strongest natural fibre available ensuring durability. Without showing the signs of wear and tear, it retains the body and lustre for years. Its abrasion resisting property has become literally proverbial. It enriches religion, tradition, ritual and culture. Since it is believed to be sacred, it has been used in ceremonies and festive; since sensuously romantic, used in matrimony; since it is property, respect, life and love, it is love and admired by the mankind. Silk with all its inherent qualities, assumes added importance when blended with other fibres. It is a fibre used in garments, tennis racket stings, fly-fishing lines, parachutes, racing cycle tyres, operation sutures and even in electric and atomic appliances. In toto silk is pronounced as the *queen of textiles* because of its beauty, prestige, practicality, safety and versatility. India loves silk because :

silk symbolises peace,

silk stands for truth,

silk stands for goodness and finally

silk stands for fairness,

Thus, silk is spectacular, majestic, magnificent as well as fascinating.

Sericulture, the cultivation of silk, is an agro-industry, the end product of which is silk. India is the second largest producer of silk in the world producing all the four varieties of commercial silk : mulberry, tasar, eri and muga. Sericulture has existed in

India for many centuries and it has a great tradition as a ceremonial robe. It is highly a viable industry which can generate vast employment opportunity in the villages. Of late, silk cultivation in India has turned out to be a highly remunerative cash crop with minimal investment but rich dividends. Sericulture has been involved in the thoughtful exercise of restructuring the country's economy towards building an economically stable India. Undoubtedly, it plays a pivotal role in effacing unemployment and in assisting the poorest to meet their two ends. Today sericulture is the main source of livelihood either directly or indirectly to a big number of people, majority of whom are from economically retarded section of the society—the Scheduled Caste (SC) and Scheduled Tribe (ST). The culture provides employment at various levels right from food plant cultivation, rearing, reeling or spinning, weaving to garment production having much impact on the improvement of rural economy of self-employment.

Silk provides much needed work and engagement in several developing and labour rich countries. It is such a product exclusively prepared by the poorest of the poor but luxuriously used by the rich. China is now the biggest silk producing country in the world where about 20 million people are engaged in sericulture and India ranking second provides job to about 5.5 million people. Though man-made fibres such as nylon, rayon, terylene, polyesters etc. continue to play their metamorphic role in the world consumption of textiles, silk, as a monomorphic fibre can supersede the demand for any modern textile and cannot be displaced from its premier position as the *queen of textiles* because of its inimitably elegant features.

India has the unique distinction of being the only country in the world bestowed by nature with all the four varieties of silk. Tasar, eri and muga culture as *non-mulberry sericulture* is known as *wild sericulture* since centuries, which has been as exclusive craft of hill folk and aboriginal like Ho, Santhal, Oraon, Munda, Kol, Koya, Puran, Bathydi, Maina, Momin, Khadia, Budha, Telaga, Nayak and other retarded sections of the community. India is the only country in the world raising all the three varieties of commercial non-mulberry silks on a large scale. The non-mulberry sericulture presents are of the finest facets of India's rich tribal culture and commercial potential but it was left almost unharnessed till the country attained freedom. The establishment of Central Tasar Research Station by Central Silk Board (CSB) at Ranchi in 1964 gave a tremendous boost to the programme of organising non-mulberry sericulture on a scientific surface. Non-mulberry sericulture has often been ignored at the international level mainly because of the erroneous notion that its potentials were too limited to be commercially viable on international scale. Orissan sericulture and tasar culture in particular today presents the picture of a well knit agrobased cottage industry firmly rooted in antiquity and traditions, with its muscles flexed to achieve continued progress through modernity and improved technology. This has been achieved through systematic planning, progressive governmental policies and hard work put in by millions of sericulturists, silk weavers and exporters all over the country during the last three decades.

Tropical and temperature tasar, eri, muga and anaphe are the principal non-mulberry silks ; the other varieties - fagara, coan, mussel and spider - are of limited interest because of less amount of silk production. All but mussel and spider silks are of insect origin. Mussel silk is obtained from a bivalve, *Pinna squamosa* found in the shallow waters along the Italian and Dalmatian shores of the Adriatic. The strong brown filament or byssus thread is secreted by the mussel to anchor it to rocky or other surface. The byssus is combed and then spun into a silk popularly known as *fish wool*. Its production is exclusively confined to Italy. Spider silk, another non-insect variety, is soft and fine but strong and elastic. The commercial production of this silk comes from certain Madagascar species including *Nephila madagascarensis, Miranda aurentia* and *Epeira*. As the spinning tubes are in the fourth and fifth abdominal segments, about a dozen individuals are confined by their abdominal part to a frame from which the accumulated fibre is reeled out four or five times a month. Because of high cost of the production, spider silk is not used in the textile industry. However, durability and resistance to extreme temperature and humidity make it indispensable for cross hairs in optical instruments.

Of all the varieties of silk fibres, nearly 95 per cent of the global production of non-mulberry silks is tasar. In India, tasar production is more than 55 per cent in comparison to eri and muga (Fig. 1.1, 1.2). Other varieties of non-mulberry silks such as eri, muga and anaphe are not commercially exploitable to a significant quantum. The small production of these non-commercially exploitable silks is mostly consumed in the producing countries. As some of the leading mulberry silk countries appear to have reached saturation point attributable to the acute scarcity of labour and the increasing cost of production in the developed countries, there is an opportunity for the developing countries like India to contribute a good amount of raw silk annually to the world's production. Precisely for this reason, the untapped and highly promising field of non-mulberry silk, the tasar has drawn the attention of rearers, users and biologists in particular.

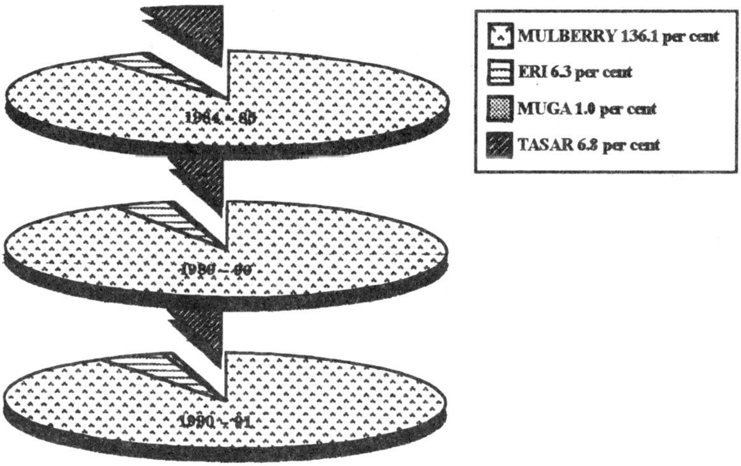

Fig. 1.1 : Raw Silk Production in India

Food plants and manpower are the basic requisites for silk production. In mulberry sericulture over 60 per cent of the cost of production goes into raising and maintaining the food plants; besides a heavy initial investment is necessary for rearing houses, rearing appliances and other essentials. Likewise, among the non-mulberry varieties eri has the disadvantage of higher production cost because it is made from domesticated silk insects. Muga silk, the monopoly of an isolated pocket of India that is Assam, is unfavourable to grow in other parts of India.

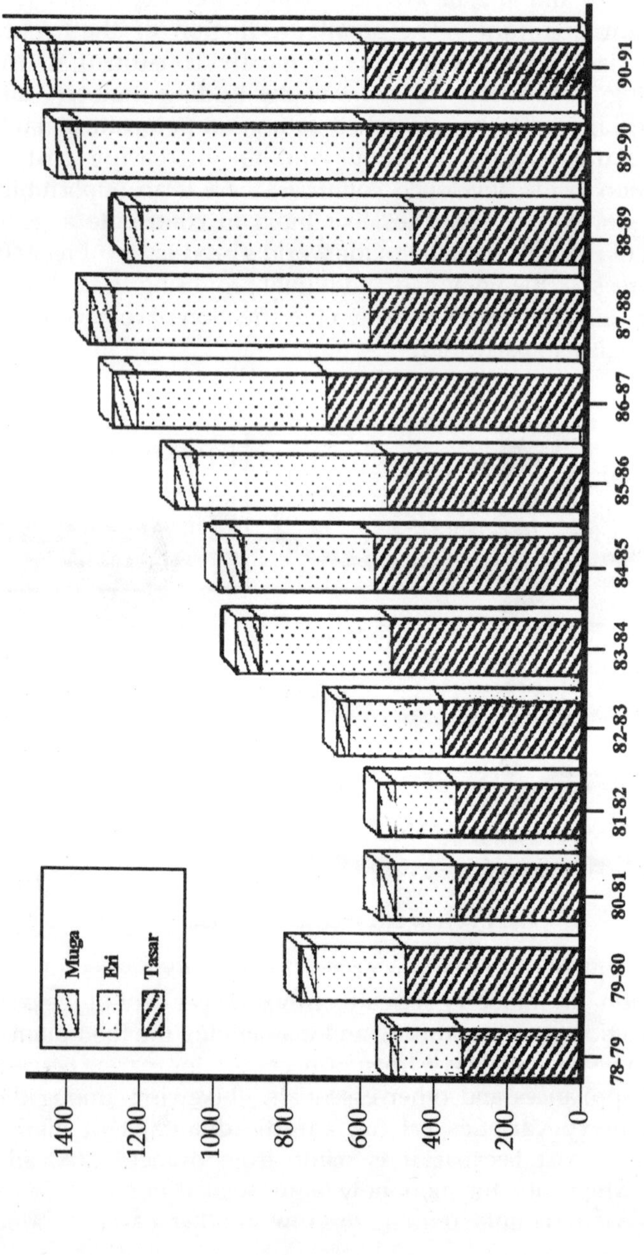

Fig. 1.2 : Non-mulberry Raw Silk Production (in tonnes)

Tasar, instead, has been endowed by nature with vast potential. The challenge before the sericulturists is to utilise this potential systematically and scientifically so as to bring about a balanced development of the silk industry in general and the rural economies of the developing nations in particular.

Thirdly, tasar cultivation holds great promise for the world forestry as a supplementary activity. On one hand it can help arrest forest destruction, and on the other it permits gainful utilisation of this vast natural wealth. Tasar culture, the most important and challenging non-mulberry silk cultures, remained obscure for centuries in the hands of the tribal population. This branch of sericulture had never been accorded any importance as in India, the annual income from tasar rearing per family was meagre. Of late, the concept of tasar culture has been radically transformed through discontinuous and unsystematic research into a viable and remunerative industry of great importance.

Fourthly, forests are an integral part of human life. These constitute not only the basic means of livelihood of millions of people but also serve, for example, to regulate precipitation, conserve soil fertility, reduce erosion and on the whole manage the ecological balance. Continuous deforestation since the beginning of the nineteenth century to meet the demands of industry, the wood trade and urbanisation is consequently a matter of grave concern. Every year more than 10 million hectares of the world's tropical forest is destroyed. There is danger that the short-term-benefit may override the long-term-disadvantages, especially to the indigenous population of forest regions. The non-mulberry tasar culture holds great promise from the world forestry as a supplementary activity.

Owing to a prolific wealth of food plants, abundant manpower and ideal climatic conditions, tasar culture offers India and Orissa in particular a unique opportunity to boost her silk production and the economic status of the tribal population. This cultivation is totally a forest based industry uniquely suited to the economy and social structure of developing countries because of its minimum investment requirement, high employment and foreign exchange earning potential. Furthermore, the various operations of tasar production can be carried out even in remote forest villages as they do not require electricity, complex machinery, specialised skills and enough time. Since tasar production offers rural population an attractive source of income, it helps arrest their migration to urban areas, thereby preserving their traditional skill and way of life. On the other hand, it simultaneously permits conservation of forest flora and fauna and economic exploitation of forest.

1.2. History of Sericulture

Silk has an obscure but romantic history. Although silk is referred in ancient Indian Sanskrit literatures, but the popular Chinese legend is treated as the most authentic one. The unravelling of silk secret is attributed to the 14 years old Chinese empress Xi Ling Shi who in 2640 B.C. unwind a glistening thread from the softened cocoon dropped into a hot tea pot. Due to her dedication and interest, sericulture began in China and she is still remembered as the *Goddess of the silkworms*. The earliest records in the chronicles of Chou-King of China in 2200 B.C. show the prominent role of silk in public ceremonies as a symbol of homage to emperors. Archaeological data on silk picture that the Chinese were using intricate designs during 1300 B.C. which were far more beautiful than anything made in recent years. In Hang Zhou summer palace, the oldest silk, estimated to be 4500 years old, are on view even today (Periasamy 1986). It is also known that a Chinese brought silkworm eggs to India when Chaui Emperor was ruling over Japan. The oldest record of sericulture is found in China written in 2600 B.C., but the silk insect eggs were first imported to Japan in 195 A.D. (Tazima 1958).

The first country after China to learn the secret of the silk was Korea, where Chinese immigrants started sericulture by about 1200 B.C. The industry later spread to Japan during 300 B.C. when Semiramus, a General in the Army of empress *Singu - Kongo,* invaded and conquered Korea. Among his prisoners were some sericulturists whom he took to Japan. Later this industry is said to have spread to India through Tibet when a Chinese princess, carrying silk insect eggs and mulberry tree seeds in her head dress married the king of Khotan in Tibet by about 1400 B.C. (Devasurappa 1986).

1.3. Silk Road

Mulberry silk spreads over through the caravans trudged past over 4000 miles long silk road which was a perilous network of routes, one along the border of the Caspian sea and other through Kashmir via Gandara on the modern North Western Frontier side. The route was hazardous to monks and pilgrims carrying Buddhist teaching between India and China and even more hazardous to traders who exchanged gold, wool, horses, jade and glass for silk.

The silk-road (Fig. 1.3) started in what is now X Ian in Shaanxi province in China, traversed a barren crust of earth through treacherous mountains and desert across Central Asia to Antioch and Tyre then by water to Europe, Egypt and other Mediterranean parts. Silk loaded caravans were passed from trader to trader at each oasis and strongholds raising the silk price higher with each exchange. Neither civilisation, East and West, were aware of the other than the edge of the route. Yet the patterns of silk were favoured by all the different cultures along the route.

1.4. Sericulture in Other Countries

History reveals that Alexander, the great, seized treasures of coveted silk when he conquered the Persian Empire. There is no doubt that silk was quite familiar to Westerners by the beginning to Christian Era. With the dawn of two powerful civilisations namely Rome and China which reigned at each end of the silk road across a bridge of culture and commerce that peaked in the Tang Dynasty (A.D. 618-907). During the period of Julius Caesar, silk was a garment of the aristocrat and the rich. Throughout centuries that followed colourful Silk Caravans trudged their way from Asia to European countries and silk was sold against the weight of gold. History also says that the famous Venetian Marco Polo, who travelled by the silk road had brought silk back to Venice from China.

In the middle ages, the rush of Islam carried silk from the Middle East across North Africa to the Siberian Peninsula. Venetians not only traded heavily but also imported silk growers and weavers to help the pioneering of their own silk industry. By the 13th century, Italy had become the silk centre of the west

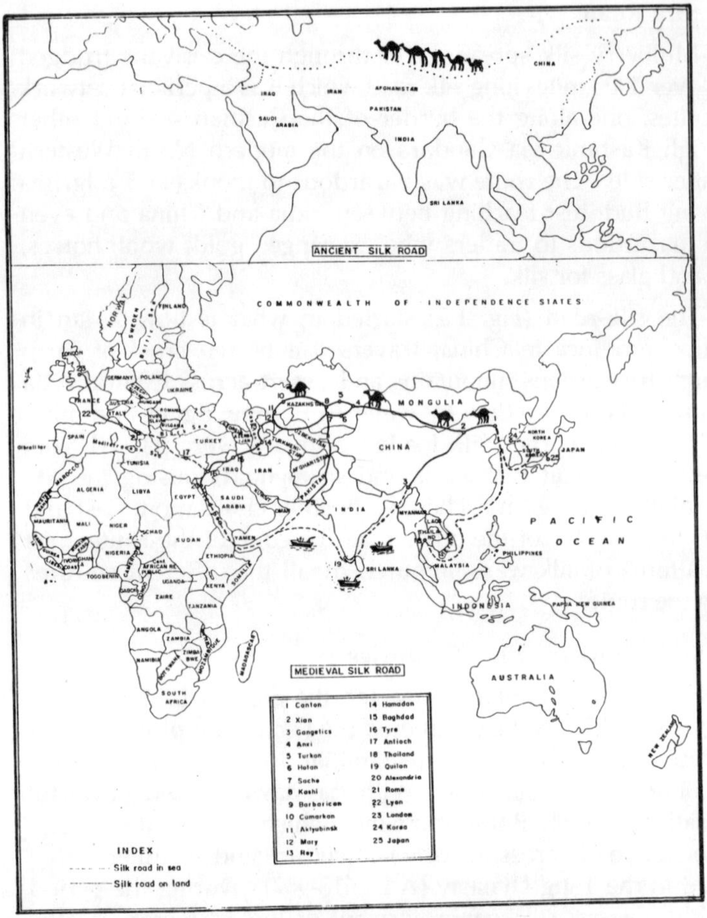

Fig. 1.3 : Silk-Road of the World

and the Italian silk industry was amazingly stabilised that helped to finance the Italian Renaissance. From Italy, the silk secret migrated to France. By 14th century *Louis XI* took drastic steps to curb the tremendous outflow of money from France to Italy and boosted France silk with royal orders. Lyon began to flourish as a silk centre in 16th century and by mid of 18th century nearly a quarter of its population were weavers. The culture touched the Americans across the Atlantic in 1522, when *Cortes* had mulberry trees eggs brought to Mexico. It grew in American colonies in 1609, when *James I* of England encouraged it to discouraged tobacco planting and fill English looms. Though the first mulberry plantings were lost at sea, ten years later, a more successful experiment in sericulture began resulting in its significant development in the colonies.

During 17th and 18th century, silk industry was favoured by the German Princess. It was regarded as a proper accompaniment of their princely dignity and an ornament of their residence towns. But before 1845, the industry died out in many places, with some vigour only at Krefeld. When the silk goods were democratised in France, Germany took advantage of this and produce silk with amazing speed during 1871. In spite of Pasteur's discoveries, silk insect diseases spread through Europe and the Middle East, which soon became increasingly dependent on the far east for raw silk. In the 18th century, the demand for silk decline during the advent of French Revolution and the industry faced a period of extreme depression. The industry, in fact, suffered considerably due to consecutive foreign invasions and absence of adequate official patronage. The industry appears to have had a prosperous time during the Mughal regime. Subsequently, it received an extraordinary stimulus during the last part of 18th century due to the effort of the East India Company with a view to securing supplies of raw silk required by the silk weaving industry of the United Kingdom. The World War II again provided a great impetus to the Indian sericulture industry. With the stoppage of supply of raw silk by Japan and China, the allied had to look to India for production supply of the strategic raw material. In 1872, the Government of Japan established sericulture station in Tomioka, Gumma Prefecture and engaged French technician to demonstrate filiature reeling. In 1884, a silkworm disease experiment station was set up which was later covered into a Sericultural Experiment Station and then into a Sericultural Training School. In 1910, the Ministry of Education established *Uyeda College of Sericulture* and silk industry.

By 1915, United States was the largest consumer of silk and Japan was the major supplier. But with the World War II, Japanese silk market was cut off and available silk was diverted to war use. In 1940s, the silk gap was partially filled up by various synthetics. Rayon and nylon hosiery quickly made a dent in silk consumption and after the war broke out silk and nylon were withdrawn from private use. In late 1970s, a better interest in the comfort and the quality of natural fibres and the demand in United States for silk clothing and have furnishings increased, which still continues as such. In 1972, Japan experienced a drastic surge in silk consumption and Japan imported silk from wherever it could get it. Western style has caused a dramatic decline in kimono wearing (typical Japanese robe) in Japan. At the same time, in order to support the silk insect growers, the Japanese Government has kept silk prices far above the world market. The modern silk industry in India has grown to meet the domestic rather than export requirements and this is a fact of great importance for the industry. The silk treasure had actually helped the renaissance of war devastated Japanese economy during post war period and helped the country to switch over to more profitable electronic trades resulting Japan to slide down to third position, China and India being first and second respectively.

1.5. Sericulture in India Through Ages

Sericulture in India is as old as the ancient Indian culture, but its real origin is obscure. It is believed to be originated from the Himalayas, Indo-Gangetic Brahmaputra flanks and in Central Plateau. Reference on silk in India are available in famous Indian Epics such as the Veda, the most ancient epic in the world, the Ramayana and the Mahabharat. In Ramayana, queen Seeta was offered a silk cloth *(Amlan Bastra)* for her entire 14 years of exile in the forest by the God Brahma, the God of creation. Silk, in fact, is a tradition and way of life in India and no ritual is believed to be complete without the use of silk in some form or other.

Western historians believe that the mulberry sericulture was brought to India from China through the famous *Silk road* via Khotan in 140 B.C. by a Chinese princess who smuggled the silkworm egg and mulberry seed to Khotan.

In Mauryan period, India has a silk trade with China during 4th to 2nd century B.C. Historians opine that raw silk was exported from India to Rome during the reign of *Kanishka* in 58 B.C. Semi-transparent silks and muslins of extreme thinness are also clearly depicted in Indian sculptures. Silk was also highly admired in the court of *King Harsha* (606-648 A.D.). Silk culture was widely practised in India during mediaeval period. It had a prosperous period during the Moghul regime particularly the Kashmir shawls and woven silks were admired in the court of Akbar.

Indian silk was widely spread out to the western countries by the British East India Company from the last part of 18th century with a view to supply raw silk requirement of U.K. *Tipu Sultan,* the ruler of Mysore, firmly established sericulture in Mysore in 1785 by importing mulberry cuttings from Bengal and silkworm seed from China. In 1869, *Maharaja Ranbir Singh* developed sericulture in Kashimr. Gradually this practice flourished in Bengal, Bihar, Assam, Kashmir and Mysore and the monopoly silk business of East India Company was abolished in 1830 and the whole trade was left to private enterprises. The silk production decreased from 1875 due to ravages of silk insect diseases and entry of Japan and China into world market. Silk industry received a brief revival during World War I. Again in 1922, the industry suffered a serious set back due to silk insect diseases, unregulated import of raw silk, lack of official support and competition from other cash crops. But in 1932, the silk industry was protected by the Tariff Board but the imperial Government did not take any planned developmental steps.

The second World War provided a great impetus to the Indian sericulture when China and Japan stopped the supply of raw silk to the allies. So innovative schemes were launched to rejuvenate sericulture in Jammu and Kashmir, West Bengal and Karnataka. Although this silk was not only of great demand for manufacturing parachutes but also for the manufacture of silk designs of Benaras and Murshidabad and its supply throughout the world. Again the industry passed through a depression period during post war period due to invention of rayon and nylon but now the real look and demand is towards silk.

In 1945, the imperial Government set up a *Silk Development Directorate.* A panel was established to suggest ways and means for the development of sericulture and accordingly the *Central Silk Board* (C S B) was established on 9.4.1949 upon an act of the Parliament of free India in 1948. The progress of sericulture during 1950-70 was slow but steady, but it is rapidly growing during the last two decades. *Central Research and Extension Stations* were established and various schemes were implemented at different states. Consequently advanced technological know how was brought from U S S R, Japan and China. The import and export was regulated and developmental steps were taken in different plan periods as a result of which sericulture took a rapid momentum. Now Indian silk is known all over the world. The masterly brocades of Benaras and Surat, soft - sigh silk of Karnataka, tie and dye patola of Gujarat, Sambalpuri of Orissa, delicate shawls of Kashmir, brilliant fabrics of Bandhej and temple silk of Kanchipuram and Tanjore are the supreme creations of mankind with a great demand throughout the globe.

2
Silk Scenario

2.1. Status of Sericulture

2.1.1. Global Status

The sericulture belt of the world runs parallel to the Tropic of Cancer up to 50° N latitude. The temperate zone covers univoltine (single crop in a year) silk producing countries such as Japan, China, Korea, Northern India, Burma, Iran, Turkey, South Russia, Lebanon, Syria, Cyprus, Greece, Rumania, Bulgaria, Hungary, Yugoslavia, Spain, Italy, France and Poland. The tropical zone extends to 10° S latitude and contains multivoltine (three crops in a year) and wild silk producing countries like South China, Central and South India, Thailand, Central Africa, Brazil and Peru. Silk is one of the world's most ancient and rarest textile fibres accounting for a mere 0.18 per cent of total textile fibre production (Table 2.1). From its halcyon days, when it was used as a form of currency, silk has gone through many vicissitudes and at one time with the advent of synthetic fibres even appeared doomed to disappear from the scene altogether. Today, silk is still very alive and kicking and world production is still at the level it was before World War II. Not only that, the production of silk is significantly increasing (Table 2.2).

In course of time, considerable change has taken place among the world's major producers. Countries such as France, Italy, Spain and Portugal have disappeared from the list of producers. Japan, the erstwhile top producer, has now declined to the status of a major producer and same is the case of the Republic of Korea. On the other hand, China has emerged as the foremost producer of the raw silk and her share of the total world production is 60 per cent. At the same time, countries as new producers have come up successfully, such as Brazil, which has become world's 2nd largest exporter of raw silk (Table 2.3). Now there are over 50 countries producing raw silk in the world, some of them with a century old tradition such as Thailand and some are new as Zimbabwe. India, which is a traditional silk producer, attempts to increase the site production and it ranks second in world as a producer of raw silk (Fig. 2.1). History establishes that as silk producing countries have become more industrialised, their sericulture industry gradually wilted and perished. This is due to competition from more lucrative activities in agriculture or industry and because of price competition from low wage paying silk producing countries (Ronald 1988).

Table 2.1 : Global production of textile fibres (Lakh tonnes)

Year	*Cotton*	*Wool*	*Silk*	*Sub-Total*	*Man-made Fibre*	*Grand Total*
1945	79.98	8.72	0.12	88.82	0.02	88.84
1950	81.02	10.44	0.19	91.65	11.31	102.96
1955	83.92	12.16	0.22	96.30	38.02	134.32
1960	95.61	13.61	0.26	109.48	44.62	154.10
1965	113.90	14.85	0.32	129.07	53.60	182.67
1970	117.82	16.02	0.41	134.25	86.36	220.61
1975	117.59	15.27	0.49	133.35	103.12	236.47
1980	142.66	15.81	0.55	159.02	132.37	291.39
1985	175.40	16.76	0.56	192.72	151.15	343.87
1990*	188.37	18.18	0.71	207.26	186.98	394.24

* Estimated

Source : Silk in India, Statistical Biennial, 1988.

Table 2.2 : Global production of silk cocoon ('000 tonnes)

Year	*China*	*India*	*Japan*	*USSR*	*Korea*	*Brazil*	*Italy*	*Others*	*Total*
1938	65.50	9.40	282.21	22.34	21.89	0.40	9.49	16.24	427.47
1945	37.38	12.11	32.65	12.88	23.33	0.20	5.52	4.16	128.23
1950	43.00	14.95	80.42	20.45	29.68	0.31	6.80	5.82	201.43
1955	111.12	24.96	104.26	25.39	6.74	0.40	8.06	6.11	287.04
1960	117.20	26.84	108.29	33.86	5.29	0.55	4.39	8.13	304.55
1965	140.00	32.99	105.51	33.00	7.76	1.56	3.33	12.78	336.93
1970	155.90	43.18	111.71	33.64	21.41	2.05	1.53	16.97	386.39
1975	199.00	44.34	91.32	38.36	36.18	7.25	0.98	19.43	436.86
1980	326.00	64.21	73.06	48.91	20.03	8.80	0.10	18.49	559.60
1985	340.00	86.18	47.26	50.61	10.23	11.84	—	31.49	577.61
1990*	470.00	115.00	28.20	56.90	8.20	15.15	—	50.50	743.95

* Estimated

Source : Silk in India, Statistical Biennial, 1988.

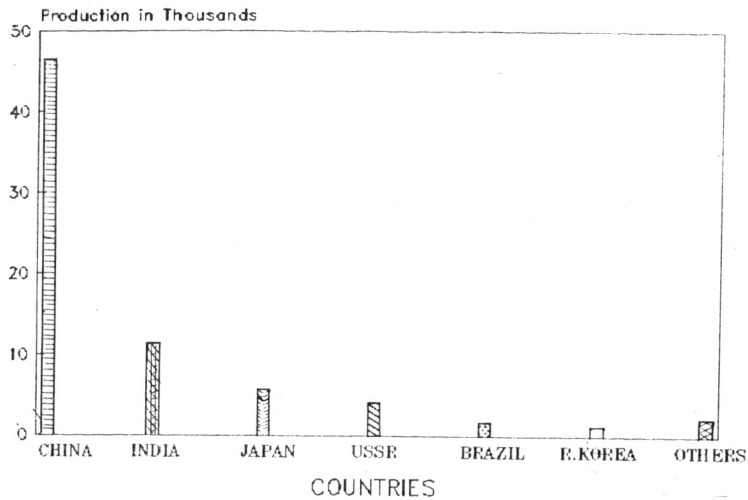

Fig. 2.1 : World Silk Production 1990

The major markets for silk products are USA, Europe, Japan and India. The demand for silk and silk materials is increasing in developed countries and expected to increase in developing

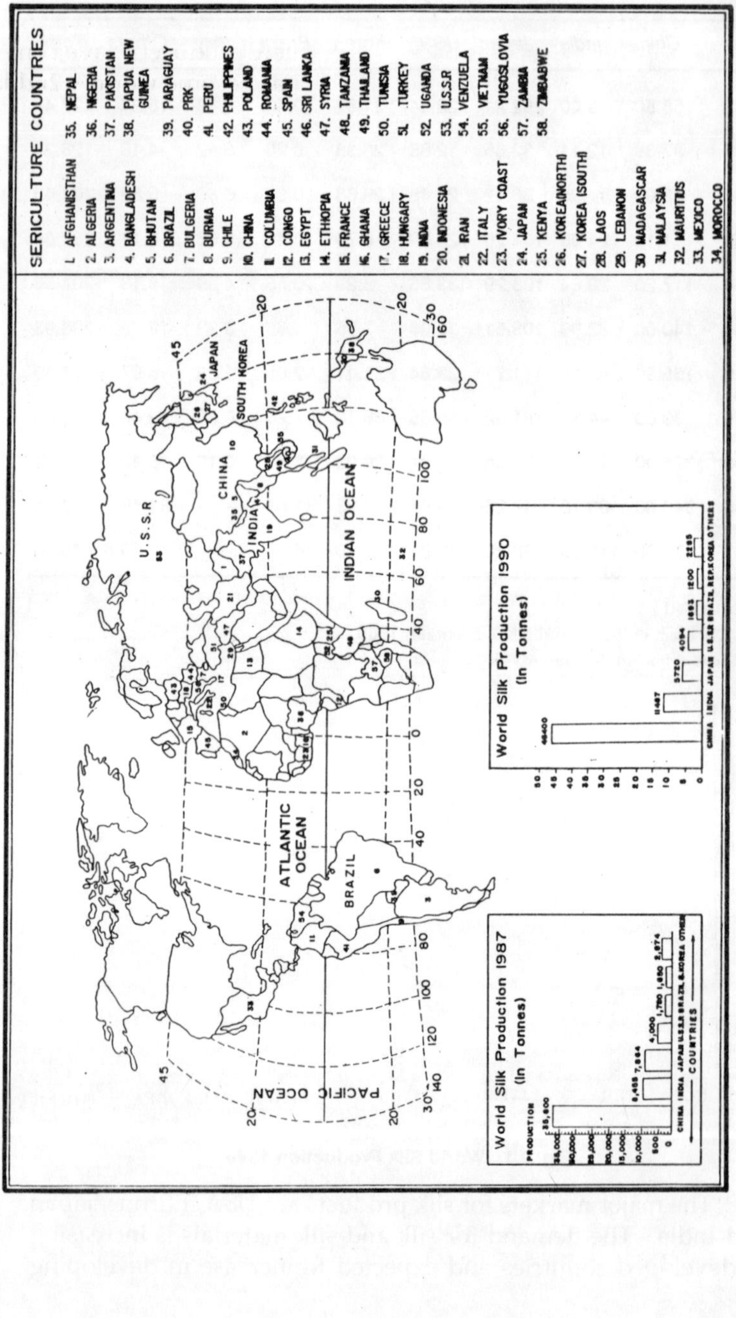

Fig. 2.2 : Sericultural Map of World

countries. The rate of production is slowly increasing and the number of countries interested in sericulture practice are coming up which is really a good sign for the growth of sericulture (Fig. 2.2).

2.1.2. Status of Sericulture in India

India has the unique distinction of being the only country in the world producing all the four types of natural silk : Mulberry, Tasar, Eri and Muga. India occupies second position in the world next to China in mulberry and temperate tasar production and first position in tropical tasar and muga silk production. Muga, the golden yellow silk, is exclusively the monopoly of India. 90 per cent of mulberry sericulture is confined to the states like Karnataka, Andhra Pradesh, Tamil Nadu, West Bengal, Jammu and Kashmir and very recently it has been introduced in Orissa, Maharashtra, Gujarat, Uttar Pradesh, Rajasthan, Bihar and Kerala. Out of the total silk production, 96 per cent is multivoltine and 3 per cent is of bivoltine variety. Tropical tasar exists in Bihar, Orissa, Madhya Pradesh, Andhra Pradesh, West Bengal, Maharashtra whereas temperate tasar is practised in Manipur, Nagaland, Mizoram, Meghalaya and Arunachal Pradesh. Eri is produced in Assam, Bihar, Meghalaya, Manipur, Arunachal Pradesh, Mizoram, Nagaland and Orissa. Muga silk is mostly confined to Assam and in small scale to Mizoram (Fig. 2.3).

India, having a total geographical area of 3.8 million square miles with a population of 846,302,688 people (Census of India, 1991), stands second among the highly populated countries in the world after China. The population of the country is estimated to reach a level of 986 million by the end of 2000 A.D. and its distribution among rural and urban areas is likely to be in the order of 660 and 326 million respectively. The rural population of our country are chiefly agricultural labourers and they account for around 67 per cent of the rural population. India has about 168 million hectares of arable land. In the post-independence period, green revolution was started in rural areas to provide employment and food to the growing population. In urban areas, due to the growth of industries, more employment opportunities were created resulting in an outflow of rural people to urban areas.

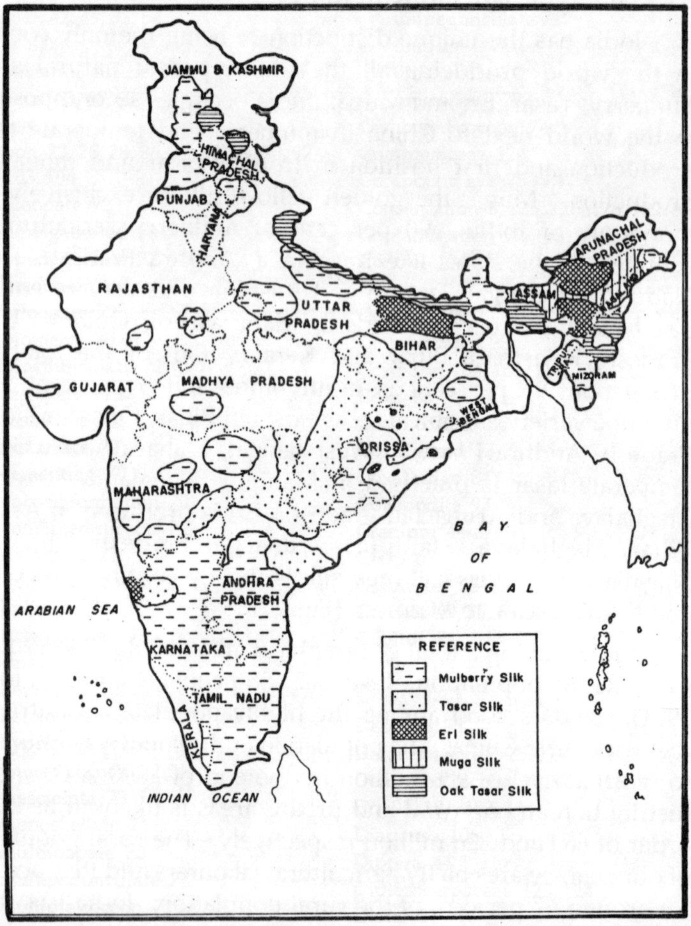

Fig. 2.3 : Sericultural Map of India

Table 2.3 : Global production of raw silk ('000 tonnes)

Year	*China*	*India*	*Japan*	*USSR*	*Korea*	*Brazil*	*Italy*	*Others*	*Total*
1938	4.85	0.69	43.15	1.90	1.82	0.03	2.74	0.66	55.84
1945	3.66	0.74	4.35	1.12	1.94	0.02	1.10	0.17	13.10
1950	4.21	0.85	10.62	1.60	1.72	0.04	1.33	0.32	20.69
1955	5.38	1.52	17.37	2.18	0.53	0.05	1.55	0.46	29.04
1960	7.58	1.61	18.05	2.90	0.47	0.07	0.89	0.54	32.11
1965	6.86	2.15	19.11	2.83	0.85	0.09	0.61	0.68	33.18
1970	11.12	2.84	20.52	3.00	3.03	0.26	0.31	1.13	42.21
1975	15.00	3.06	20.17	3.00	5.55	0.93	0.14	1.68	49.53
1980	23.48	5.04	16.16	4.25	3.33	1.28	0.01	3.14	56.69
1985	32.00	7.89	9.59	4.00	1.85	1.56	—	3.17	30.06
1990*	45.00	10.50	5.50	4.50	1.50	2.00	—	5.00	74.00

* Estimated

Source : Silk in India, Statistical Biennial, 1988.

Sericulture being a labour intensive agro-industry offers a great advantage of stabilising the rural population in the country avoiding mass exodus towards overcrowded cities and at the same time provides a sound source of income. Among cottage and village industries, sericulture is the largest employment after handlooms. Sericulture provides gainful employment to small and marginal farmers and particularly SC and ST. It is a highly remunerative cash crop with rich dividend in minimum investment of labour (Fig. 2.4). Silk industry plays a pivotal role in transferring wealth from the rich to the poor as silk is consumed mostly by the affluent. The money so spent by them on purchase of silk is distributed among different group of people engaged in it (Table 2.4). Now sericulture has spread over 59,528 villages out of 6,29,143 villages of different states of the country (Table 2.5). Realising the importance of silk cultivation in rural economy, both Central and State Governments emphasise

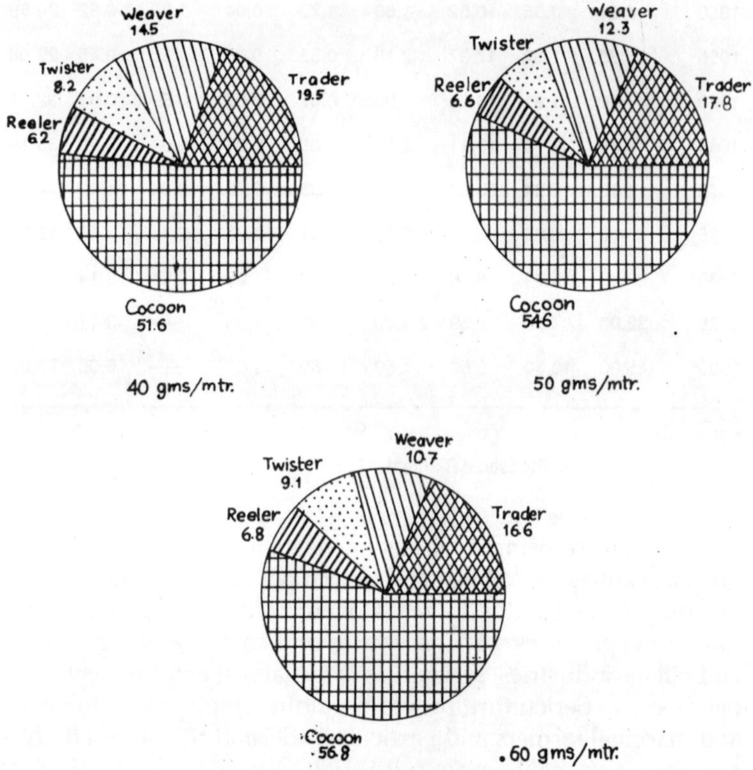

Fig. 2.4 : Distribution of Gross Income from Soft Silk Fabrics

on sericulture as an effective tool of rural development and employment generation.

Table 2.4. : Earning distribution in sericulture

(% Share)

Category of persons	*Soft silk fabric of* 40 gms/Mtr	50 gms/Mtr	60 gms/Mtr.
Cocoon producer	51.6	54.6	56.8
Reeler	6.2	6.6	6.8
Twister	8.2	8.7	9.1
Weaver	14.5	12.3	10.7
Trader	19.5	17.8	16.6
Total	100.0	100.0	100.0

Source: 1. Silkman's Companion, CSB, 1989.
2. Silk in India, Statistical Biennial, CSB, 1992.

2.1.3. Importance of Sericulture in India

At present three lakh hectares of land are available under mulberry silk cultivation producing about 90,508 million tonnes of cocoons and 9,350 million tonnes of raw silk. Production of non-mulberry silk is to the tune of over 1,154 metric tonnes which includes 617 metric tonnes of tasar, 465 metric tonnes of eri and 72 metric tonnes of muga (Table 2.6). In India, silk reeling capacity consists of 21,983 charakhas ; 6,955 cottage basins and 8,879 filiatures with 2 lakh spindles (Silk in India, CSB 1988). The rate of production by charakha, cottage basin and filiature is 50 per cent, 42 per cent and 8 per cent respectively. Five spun silk mills are also established for spun silk production and noil yarn from silk waste. At present, there are about 23,600 spindles for spun silk and 3,462 spindles for noil yarn (Silk in India, CSB 1992).

Table 2.5 : Sericultural zones in India (1989-90)

Sl. No.	*State*	*Total number of villages*	*No. of sericultural villages*	*Percentage to Total*
1.	Andhra Pradesh	29,281	9,377	32.0
2.	Arunachal Pradesh	3,257	100	3.1
3.	Assam	23,919	7,102	29.7
4.	Bihar	77,848	4,777	6.1
5.	Gujarat	18,550	137	0.7
6.	Himachal Pradesh	18,721	1,687	9.0
7.	Jammu & Kashmir	6,758	2,620	38.8
8.	Karnataka	29,390	16,593	56.5
9.	Kerala	1,219	—	—
10.	Madhya Pradesh	76,468	1,314	1.7
11.	Maharashtra	41,833	1,357	3.2
12.	Manipur	2,059	354	17.2
13.	Meghalaya	5,048	1,812	35.9
14.	Mizoram	736	60	8.2
15.	Nagaland	1,119	140	12.5
16.	Orissa	50,887	2,300	4.5
17.	Punjab	12,822	818	6.4
18.	Rajasthan	37,124	114	0.3
19.	Sikkim	447	6	1.3
20.	Tamil Nadu	16,812	4,985	29.7
21.	Tripura	864	278	32.2
22.	Uttar Pradesh	1,24,246	2,017	1.6
23.	West Bengal	41,107	1,580	3.8
	Total	6,29,143	59,528	9.5

Source : Silk in India, Statistical Biennial, CSB, 1992.

Table 2.6 : Silk scenario in India

	COCOON				RAW SILK (in tonnes)						SILK WASTE (in tonnes)					
	Mulberry*	Non-mulberry			Mulberry	Non-mulberry					Mulberry	Non-mulberry				
		Tasar**	Eri***	Muga****		Tasar	Eri	Muga	Sub total	Total		Tasar	Eri	Muga	Sub total	Total
1938	9.40	68	129	10	575	81	28	10	119	694	424	44	14	4	62	486
1945	9.21	88	132	12	561	89	71	21	181	742	285	63	22	11	96	381
1950	11.43	139	161	15	610	115	92	36	243	853	315	74	30	14	118	433
1955	19.69	142	268	25	1087	228	161	48	437	1524	452	102	52	17	171	623
1960	20.83	145	255	25	1154	242	170	52	464	1618	581	133	84	22	239	820
1965	25.04	217	329	26	1634	257	203	58	515	2149	714	181	63	28	272	986
1970	34.28	225	274	23	2258	353	164	69	586	2844	939	167	53	28	248	1187
1975	36.74	237	177	2	2541	360	123	43	526	3067	1037	158	43	13	214	1251
1980	58.21	246	200	11	4593	265	135	48	448	5041	1376	99	8	29	136	1512
1985	76.72	468	555	27	7029	464	352	52	868	7897	2504	130	32	13	175	2679
1990	90.50	520	850	45	9350	617	465	72	1154	10504	3330	170	42	18	230	3560

* '000 Tonnes

** '000 Kahan (1 Kahan = 1,600 nos.)

*** Tonne

**** Crore number

Silk weaving is largely undertaken on handlooms whose strength is 1,87,452 producing 94 per cent of the fabric. Remaining 6 per cent of silk weaving is practised on 8,212 powerlooms (Silkman's Companion, CSB 1989). However, the total fabric production in the country is more than 85 million square metre. The domestic market for silk is essentially sarees, which is about 90 per cent. The size of Indian domestic silk market is estimated to be worth Rs. 2,000 crores involving 3,500 metric tonnes of raw silk. The average consumption growth rate is about 12 per cent (Silkman's Companion, CSB, 1992). A steady growth in the export of Indian silk goods is over 25 per cent per annum. The main items of export are silk fabrics, readymade garments and carpets. Now the size of the silk export is more than 250 crores. To regulate the silk export, Silk Export Promotion Council has been established. A small quantity of raw silk (100 metric tonnes) is imported for use in exportable silk fabrics and to meet the special requirement.

Cocoon transactions have been done through regulated co-operative markets operated by the Government and linked to the Central Raw Material Bank. Silk exchanges are established for the proper sale of raw silk. The market price of silk and silk material is regulated by the Silk Marketing Board. CSB supervises all the research and development (R and D) projects in this country. Sericulture development, both under mulberry and non-mulberry sector, was handicapped for a considerable period due to low productivity, poor quality, high cost of production and lack of appropriate technologies. Establishment and reorganisation of research and training institutes have solved these problems to a great extent and attempt is going on for better productivity.

2.1.4. Status of Sericulture in Orissa

Orissa is one of the important silk producing states in India along with other states. Sericulture activities in Orissa comprise of rearing of three varieties of silkmoth namely Tasar, Eri, Mulberry. These cocoons are produced to silk yarn and utilised for the production of silk fabrics. Tasar silk insect rearing is a traditional occupation of 50,000 tribals in Orissa since time immemorial. Tasar culture is dominated in the district of Mayurbhanj, Keonjhar and Sundargarh. There are also some

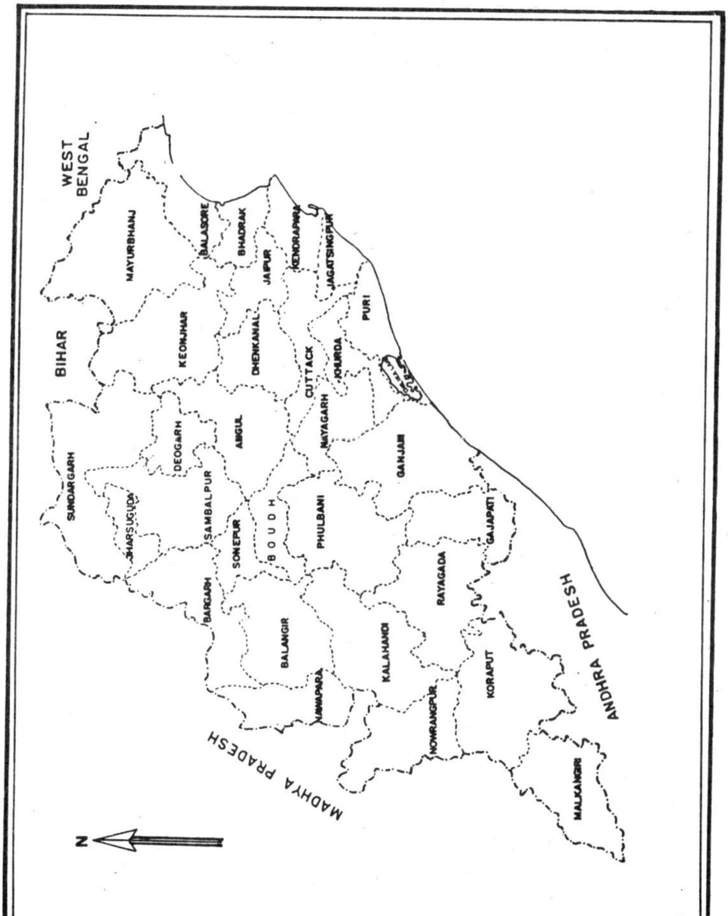

Fig. 2.5 : Tasar Cultivated Zones of Orissa

pockets of intensive rearing of tasar moth in the districts of Balasore, Cuttack, Dhenkanal, Kalahandi, Sambalpur, Bolangir, Phulbani, Ganjam and Koraput (Fig. 2.5). Average cocoon production in the state is 75,000 kahan (1 kahan = 1600 cocoons in Orissa).

Mulberry sericulture was introduced in the state since 1977 on an experimental basis and now it has been extended to 40 selected blocks having suitable agroclimatic condition. Average production of mulberry cocoons in the state is 25,000 kg involving around 4,000 people. Eri silkworm rearing was introduced in Orissa during the year 1942 in the district of Cuttack, Puri and Ganjam (Coastal belt of Orissa) but subsequently abolished from these places. Now it is extended into the tribal dominated districts such as Phulbani, Keonjhar, Sambalpur and Koraput involving about 2,000 people. The average production of eri cocoon in Orissa is 2,000 kg. It is observed that the sericulture activities in Orissa is mostly confined to tasar culture. Tasar cultivation is a wild culture traditionally practised in the remote forest areas in a very crude way by the primitive tribal people from time immemorial which needs systematic survey and research for its allround development and better economic utilisation.

2.1.5. Types of Silk

The natural silks are broadly of two types : mulberry and non-mulberry or wild silk. Mulberry silk-moth is a domesticated type whereas non-mulberry is of wild type or forest variety and semidomesticated one. Further, the non-mulberry variety of silks are classified as insect and non-insect type. Insect type of silk is obtained from five different types of insects and the silk is named as (i) Eri (ii) Muga (iii) Anaphe (iv) Fagara (v) Coan and (vi) Tasar and that from non-insects is of two types : (i) Mussel silk and (ii) Spider silk being obtained from mollusc and spider respectively (Fig. 2.6). Again the tasar is classified into two types according to distribution such as (i) Temperate and (ii) Tropical tasar (Fig. 2.7). For the entire category of silk, the host plants and the distribution of the fauna is almost variable (FAO 1987) excepting mussel silk and spider silk (Table 2.7). This table also indicates that mulberry silk moth is monophagous whereas the nonmulberry insect variety of silk-moths are usually polyphagous.

Table 2.7 : Silk variety, fauna, host plant and distribution of silk

Sl. No.	*Type*	*Common name of the silk*	*Scientific name of the animal*	*Common name of the host plant*	*Distribution of silk*
1.	Insect	Mulberry	*Bombyx mori* L.	Mulberry	Japan, Korea, U.S.S.R., India, China, France, Spain, Italy, Iran, Thailand, Bangladesh, Poland
2.	Insect	Tasar (Tropical)	*Antheraea mylitta* Drury *A. paphia* Linn.	Sal, Asan, Arjun	India
3.	Insect	Tasar (Temperate)	*A. yamamai, A. proylei* Jolly *A. pernyi* Guerin	Oak	Japan, India, China
4.	Insect	Eri or Endi	*Philosamia ricini* Boisd. *P. cynthia* Drury	Castor, Kesseru, Simulalu, Papita, Pyam, Maharukh	Assam, Bihar, West-Bengal Manipur and Tripura
5.	Insect	Muga	*Antheraea assamensis* Ww.	Som, Scalu, Chapa, Mejankori, Bhomlati	Assam
6.	Insect	Anaphe	*Anaphe moloneyi* Druce *A. panda* Boisduval	Berkoli and others (Polyphagous)	Southern and Central Africa
7.	Insect	Fagara	*Attacus atlas* L.	Polyphagous	Indo-Australian region, China and Sudan
8.	Insect	Coan	*Pachypasa otus* D. *P. lineosa* Vill.	Pine, Ash, Cypress, Juniper, Oak	S. Italy, Greece, Romania and Turkey
9.	Non-insect	Mussel	*Pinna squamosa*	———	Italy
10.	Non-insect	Spider	*Nephila madagascarensis* *Miranda aurentia* *Epeira* sp.	———	New Guinea, Papua, North Queensland

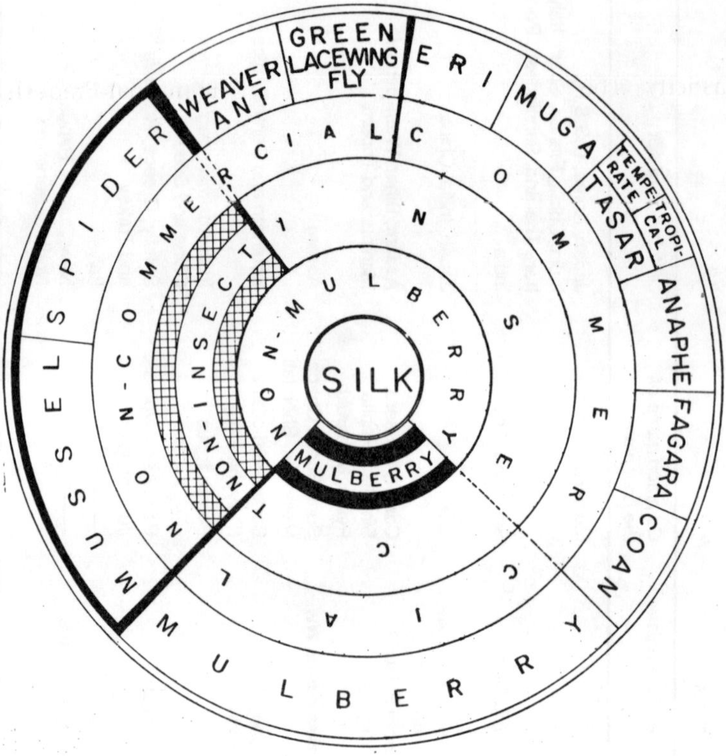

Fig. 2.6 : Varieties of Silk in Animal World

(i) Eri Silk

The white or brick red eri silk is also known as *endi* or *errandi* which is produced by *Philosamia ricini* Boisduval. This insect is a domesticated multivoltine silk-moth. Ericulture is widespread in Assam and practised on a small scale in Bihar, West Bengal, Manipur, Orissa and Tripura.

(ii) Muga Silk

The muga silk, popularly known as golden-yellow silk, is obtained from a semidomesticated multivoltine insect *Antheraea assamensis.* This muga culture is exclusively the monopoly of Assam. Larvae of muga when fed on megankori leaves (*Litsea citrata)*, produce a kind of silk known as mejankori silk which is attractive due to its durability, lustre and creamy white shaded fibre.

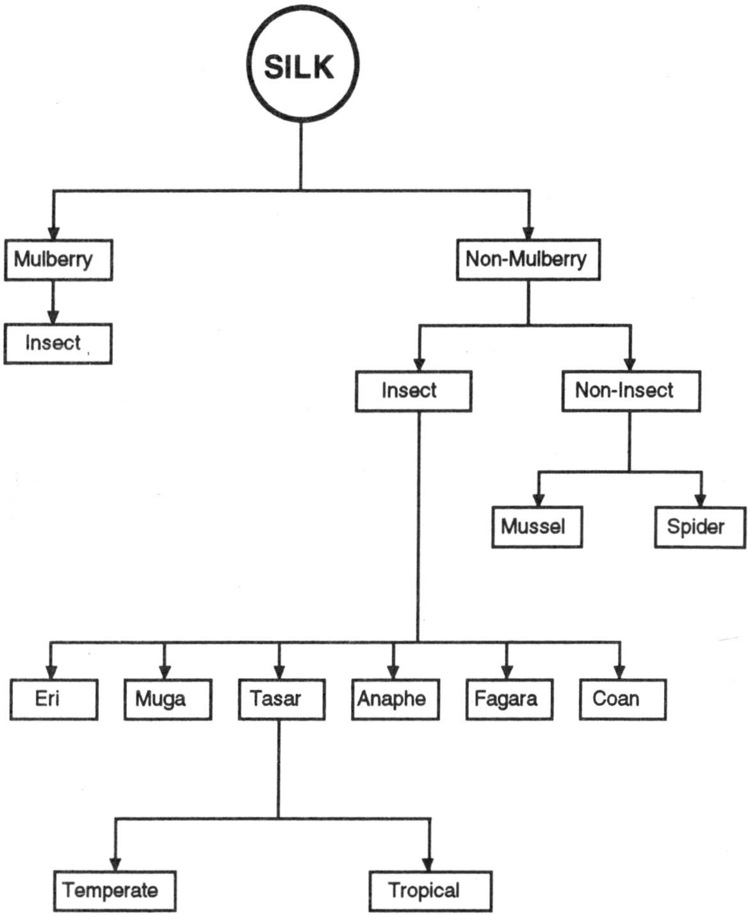

Fig. 2.7 : Schematic Classification of Silk

(iii) Anaphe Silk

The name anaphe is coined after the genus *Anaphe* and the silk is distributed in Southern and Central Africa. The species under the genus *Anaphe* are *A. moloneyi* Druce, *A. panda* Boisduval, *A. reticulata* Walker, *A. ambrizia* Butter, *A. carteri* Walsingham, *A. venata* Butter and *A. infracta* Walsingham. They spin cocoons in communes, all enclosed by a thin layer of silk. The fabric is elastic and stronger than mulberry silk.

(iv) Fagara Silk

Fagara silk is obtained from the giant silk-moth *Attacus atlas* L. and few other related species inhabiting the Indo-Australian biogeographic region, China and Sudan. The fagara cocoons are light-brown in colour and has got less importance since the silk is not commercially exploitable.

(v) Coan Silk

Coan silk fibre is secreted by the larvae of *Pachypasa otus* D. These larvae are found in the Mediterranean biogeographic region like Southern Italy, Greece, Romania and Turkey. This is a polyphagous insect feeding on pine, ash, cypress, juniper and oak. The cocoons are white in colour which was used to make the crimson - dyed apparel worn by the dignitaries of Rome. Cultivation of coan silk is dropped due to its limited out put.

(vi) Mussel Silk

Mussel silk is a non-insect type of silk obtained from a bivalve mollusc, *Pinna squamosa.* These hard shelled animals are distributed along the Italian and Dalmatian shores of the Adriatic. The fibres are called as byssus threads which are brown in colour, strong in quality and help the animal to anchor itself to a rock or any surface. The byssus is combed and spun into silk popularly known as *fish wool.* Its production is largely confined to Toronto, Italy.

(vii) Spider Silk

Spider silk is another non-insect type of silk spun by arachnids like *Nephila madagascarensis, Miranda aurentia* and *Epeira.* This silk fibre is reeled out four or five times a month from the spinning tubes located at the 4th and 5th abdominal segments of the spider body. The spider silk is derived from transformed coxal glands in the abdomen, whereas the silk of the silk insect comes from modified salivary glands located in the head. Spider silk is noted for its strength and elasticity. The considerable tensile strength which is second only to fused quartz fibres and far greater than steel, goes hand in hand with great elasticity. The strength of the threads is to some extent dependent on the manner in which the spider draws them out with greater speed increasing the strength of the thread. When they are drawn speedily, the fibroin chains attain a maximum orientation, which contributes greater strength to the lines. The cocoon silk of the silk insect is essentially equal in strength to that of orb-weaving spiders (Gertsch 1979). Spider silks were used as gill nets, dip nets, kite nets and various lures for the fishing activities and also for weaving bags, caps and head dresses by primitive people like certain New Guinea natives. In 1709, a Frenchman *Bon de Saint-Hillare,* demonstrated the use of spider silk in the same way as the silk of the silk insect. However, solitary and predaceous nature, varied thickness of the silk and low quantum of silk production are the basic obstacles of spider silk as a practical commercial textile.

(viii) Weaver-Ant Silk

Weaver-ants of the genus *Oecophylla* are found to build silk nests in trees. All these weaver-ants belong to this single genus which has two living species such as:

(i) *Oecophylla longinoda* which is commonly known as the yellow tree ant of tropical Africa and

(ii) *Oecophylla smaragdina* which is the red tree ant of South Eastern Asia and Mehanesia.

Some other less well-known ants also use silk in building nests. The silk that all these ants use is not produced by the workers but comes from the larval glands which generally provides silk for the cocoon. *Oecophylla* and species of *Polyrachis* which make silk nests do not spin cocoons at pupation, although most of the Formicinae do so (Wheeler 1915, Ledoux 1949).

Some species of *Polyrachis* use silk to reinforce the walls of the earthen cells in their nests. When larvae of species which have cocoons spin they are usually partly earthed up by workers.

This evidently gives the mature larva an anchorage for the first threads of silk and allows it to draw out threads from its gland. Perhaps the mixture of silk with earth in *P. gagates* and in young colonies of *P. laboriosa* arose from this habit. Older colonies of *P. laboriosa* extend their nests into trees where they build bags of loosely felted silk in which dead leaves are entangled. Some other *Polyrachis* species build flatter earth and silk cells on the underside of leaves (Wheeler 1915, Ledoux 1958).

Oecophylla nests are not made in this simple fashion. Instead it draws living leaves together and uses larval silk to secure them, and to seal up the chinks between them. As the leaves die the ants make new nests in fresh leaves, and they also quickly repair rents in their leaf nests. *Oecophylla* is able to make its nests from a wide variety of leaves including pinnate palms, the paired ovate leaflets of *Bauhinia* and the large stiff leaves of *Citrus*.

(ix) Mulberry Silk

Mulberry silk is widely cultivated in the globe which is believed to be a native, either of India or China. Mulberry is produced by the mulberry silk insect *Bombyx mori* L. which is monophagous, *i.e.*, feeding on the single food plant mulberry (*Morus*). The bulk of the silk goods produced in the world are from mulberry silk-moths. The climatic conditions for mulberry culture are favourable for luxuriant growth of mulberry and rearing of silk insects throughout the year. It is also highly economical since the production rate is high against a minimum expenditure.

(x) Tasar Silk

Tasar silk is another insect variety of silk secreted by *Antheraea* tasar moths. Tasar insects are polyphagous feeding on both primary and secondary food plants. *Antheraea* comprises of more species than any other genus of sericigenous insects. The master creators of the world famous Indian tasar belong to aboriginal forest peoples who have been rearing tropical tasar for centuries. The dense humid tropical forest sprawling over the Central and Southern Plateau is the habitat of tropical tasar moth. The second type of tasar is temperate tasar silk which is largely confined to the north-east provinces. The production takes place by temperate tasar silk-moth *Antheraea pernyi*. This industry has progressed rapidly, particularly in the north eastern states owing to ideal socio-economic and rearing conditions.

2.2. Significance of Tasar Culture

The term *Tasar* has been apparently derived from the Sanskrit word *trasara* meaning shuttle. Tasar is a type of silk produced by the wild silk-moths or giant silk-moths, which is also termed as *tussah, tassar* or *tussora*. Silk cultivation of tasar or tasar culture in India is not only an industry but also a tradition. It remained obscure for a pretty long time as an exclusive craft of the tribal, hill folks and aboriginal inhabiting the remote forest areas of Central Plateau and North Eastern area. It was in recent past that this tribal tradition assumed importance and attracted the attention at national level. The rich production potentials within the country and steady demand of the products in foreign countries promoted commercial exploitation of this craft and thus, culminated in the transformation of age-old tradition to an industry of immense potentiality.

Tasar culture is practised on the forest plants in wild condition. The great advantage of tasar culture is that it does not require any investment on plantation, rearing, appliances, electricity and other essentials like mulberry sericulture. Tasar cultivation not only checks destruction of forest but also provides a gainful utilisation of vast natural resources with minimum investment alongwith huge employment generation in rural areas and development of down-troddens.

Tasar, today, is an indispensable and growing part of tribal culture and medium for the transformation of the whole rural economy in relation to the national structure. The industry has rich natural resources (food plants) and man power (tribal) and the challenge to utilise these to bring about a balanced development in keeping with the tradition and way of life. The concept of tasar culture has undergone radical transformation with the recent introduction of tasar on oak trees. This practice has two distinct entities : tropical and temperate. Tropical tasar growing area forms a distinct belt of humid and dense forest sprawling over Central and Southern Plateau. It covers the principal states such as Bihar, Madhya Pradesh, Orissa, West

Bengal and touching the fringes of Uttar Pradesh, Andhra Pradesh, Maharashtra and Karnataka. Temperate tasar extends from Jammu and Kashmir in west to Manipur in east including Himachal Pradesh, Uttar Pradesh, Assam, Mizoram, Arunachal Pradesh, Meghalaya and Nagaland. Out of the total tribal population of about 30 million, nearly 77 per cent are located in tasar cultivated states of tropical (25.67 million) and temperate (4.23 million) belts. However, 12.7 and three million tribals live in tropical and temperate tasar cultivation districts respectively while 1.27 lakh tribals are engaged in tasar industry in tropical zone and 5 to 7 thousand of them in temperate belt and equal number of families are engaged in tasar weaving and reeling.

There is a vast wealth of tasar food plants both in tropical and temperate zones and it is estimated as 11.16 million hectare and 1.26 million hectares respectively. At present, only five per cent of the area of tropical food plants and hardly 0.8 per cent of the area of the temperate tasar food plants have been exploited as such. The tasar industry has the potentiality of generating huge employment opportunity and foreign exchange earning.

2.3. Tasar Silk-Moths

The Indian tropical tasar silk insect is found in wild and semi-domesticated condition. Depending upon their commercial characters and place of availability, 25 ecoraces have been grouped. In Orissa, the ecoraces of Antheraea mylitta are known as *Sukinda, Daba, Modal* and *Bogei* (*Jolly 1979*). But from the survey, it is understood that same group of cocoons available in a particular place having some definite economic characters have been classified into an ecorace and named after their local names. In fact, the wild tasar silk insects of Orissa, which reproduce in nature are identified as *Antheraea paphia*. Cocoons of *A. paphia* produced at higher altitude (601-1,000 m ASL) is locally known as *Modal* having more economic trait and that of the middle altitude (301 - 600 m ASL) is known as *Nalia* have lesser economic trait. Both Modal and Nalia when cultivated in the lower altitude (50 - 300 m ASL) by the rearers, the progeny cocoons are locally known as *Bogei* have lesser economic trait than their parents (Table 2.8).

Table 2.8 : Survey of tasar silk-moth in different altitudes

Altitude (m ASL)	*Antheraea paphia*			*Antheraea mylitta*			*Frequency of Life Cycle*
	Local name	*Voltinism**	*State of Cultivation*	*Local name*	*Voltanism**	*State of Cultivation*	
50 - 300	Bogel	TV or BV	Cultivated	Sukinda	TV	Semi-domesticated	Three
301 - 600	Nalia	BV	Wild	Daba	BV	Semi-domesticated	Two
601-1,000	Modal	UV	Wild	—	—	—	One

* TV - Trivoltine BV - Bivoltine UV - Univoltine

The semidomesticated tasar silk insects of Orissa which have been reared by the rearers through countries are identified as *Antheraea mylitta.* It is never cultivated at the highest altitude (601 - 1,000 m ASL). *A. mylitta* cultivated in the middle altitude (301 - 600 m ASL) is known as *Daba* and that of the lowest altitude (50 - 300 m ASL) is *Sukinda.* As per practice Daba reproduce twice and Sukinda thrice in a year. *Antheraea mylitta* is a polyphagous and semidomesticated tasar silk insect, feeding on the leaves of the Asan (*Terminalia tomentosa),* Arjun (*T. arjuna*) and Sal (*Shorea robusta*). Most of the ecoraces are bivoltine and a few are trivoltine (Jolly *et al.* 1974, 1979). Depending upon the voltinism of the race, the reproduction coincides with rain (July-August), autumn (September - October) and winter (November-December) seasons. In Orissa, *A. mylitta* has been reared on Asan and Arjun plants and exhibits bi and trivoltinism in two ecoraces such as *Daba* and *Sukinda* respectively.

Table 2.9 : Vegetation cover in India, 1991

Sl. No.	State/UTs	Area in sq. kms Geographical Area (Area in sq. kms.)	Actual Forest Cover	Actual Forest Cover as per cent of Geographical Area
1.	Andhra Pradesh	2,75,068	47,290	17.2
2.	Arunachal Pradesh	83,743	68,757	82.1
3.	Assam	78,438	24,751	31.5
4.	Andaman and Nicobar	8,249	7,622	92.4
5.	Bihar	1,73,877	26,668	15.3
6.	Goa	3,702	1,253	33.8
7.	Gujarat	1,96,024	11,907	6.1
8.	Chandigarh	114	5	4.5
9.	Delhi	1,493	22	1.5
10.	Haryana	44,212	513	1.2
11.	Himachal Pradesh	55,673	11,780	21.2
12.	Jammu and Kashmir	2,22,235	20,064	9.0
13.	Karnataka	1,91,791	32,199	16.8
14.	Kerala	38,863	10,292	26.5
15.	Manipur	22,327	17,685	79.2
16.	Maharashtra	3,07,690	44,044	14.3
17.	Mizoram	21,081	18,853	89.4
18.	Punjab	50,362	1,343	2.7
19.	Rajasthan	3,42,239	12,835	3.7
20.	Sikkim	7,096	3,033	42.7
21.	Tamil Nadu	1,30,058	17,713	13.6
22.	Tripura	10,486	5,535	52.8
23.	Uttar Pradesh	2,94,411	33,609	11.4
24.	West Bengal	88,752	8,015	9.0
25.	Meghalaya	22,429	15,875	70.8
26.	Nagaland	16,579	14,321	86.4
27.	Orissa	1,55,707	47,205	30.3
28.	Dadra and Nagar Haveli	491	206	42.0
29.	Daman and Diu	112	2	1.8
30.	Lakshadeep	32	—	—
31.	Pondicherry	492	—	—
32.	Madhya Pradesh	4,43,446	135,785	30.6
	Total	32,87,263	639,182	19.44

Source : The State of Forest Report, GOI, 1991.

Table 2.10 : Deterioration of natural forest and tasar food plants in India and Orissa (in million hectare)

Sl, No.	*Nature of the forest*	*INDIA*			*ORISSA*		
		1980	*1995*	*Depletion*	*1980*	*1995*	*Depletion*
1.	Total forest area	147.2	67.2	80.0	6.08	5.48	0.60
2.	Total tropical forest area	44.5	20.8	23.7	6.08	5.48	0.60
3.	Total forest in Tasar belt	11.7	5.3	6.4	2.02	0.89	1.13
4.	Total sal forest	9.7	4.8	4.9	1.69	0.77	0.92
5.	Total arjun forest	2.0	0.9	1.1	0.33	0.12	0.21
6.	Total tasar rearing area	0.4	0.2	0.2	0.051	0.024	0.027

Antheraea paphia exhibits uni and bivoltinism in two ecoraces such as *Modal* and *Bogei* respectively. *Modal* is a wild tasar silk insect grown in natural Sal forests during rain (July-August) season above 600 m ASL and is the highest silk producing insect ever known whereas *Bogei* is semidomesticated and reared on Asan plants (*T. tomentosa*) during rain and autumn seasons below 500 m ASL and produces superior cocoons than that of *A. mylitta*. However, tasar cocoon production is not in increasing quantum because of deterioration of natural forest in general (Table 2.9) and natural wealth of tasar food plants in particular (Table 2.10). Deterioration of the natural forest is the prime cause of insignificant increase of raw silk production in India.

3

Biology of Tasar Silk-Moth

Tasar silk-moth *Antheraea paphia* is a member of the super family Bombycidae of the family Saturniidae being characterised by the prominent bipectinate antennae in both sexes and a transparent eye spot at the centre of each wing. As a homometabolous insect, tasar silk-moth exhibits four distinct stages : egg, larva, pupa with cocoon and adult moth in its life cycle.

3.1. Stages in Life Cycle

3.1.1. Egg

The egg is comparatively large, oval, flattened and bilaterally symmetrical along the antero - posterior axis. It is creamy in colour but slightly bluish just before hatching. It has two dark brown lines, running parallel to each other around the periphery. There are three different axes : (i) the longest micropylar axis measures on an average 0.3 cm and is horizontal to the surface of deposition of the egg, (ii) the axis along the width measures on an average 0.25 cm and (iii) the shortest vertical axis measures on an average 0.15 cm. The egg appears to be elliptical in shape when looked down from the micropylar end with the diameters of 0.25 cm and 0.15 cm. The surface is so clouded with dark brown gum by which the eggs are firmly glued to one another and to the surface of deposition that the true creamy colour of the egg is concealed. On cleaning this substance, a smooth creamy surface is revealed with faint pentagonal and hexagonal netting and slight depressions throughout. The mircopyle is found to be situated at one end of the egg in a slightly raised area in between the two parallel lines. These lines are slightly broadened at this end and the micropyle is marked by a distinct dark spot (Fig. 3.1). The egg has a hard shell and without the colour of the serosa, as such the development of embryo is difficult to determine from the external side. Both the ends of the egg are pretty equal in size not being more narrowed towards the micropyle than it is at the opposite end. The egg measures 3.01 mm along the micropylar axis, 2.47 mm in width and 1.56 mm along the vertical axis and the average weight of an egg is 9.8 mg. Eggs deposited in later days are lighter than those laid during first two days. The weight of the eggs gradually decreases from the first day to sixth day. The average weight of the first day eggs is 10.34 mg and that of sixth day is 9.23 mg showing a successive decrease in weight.

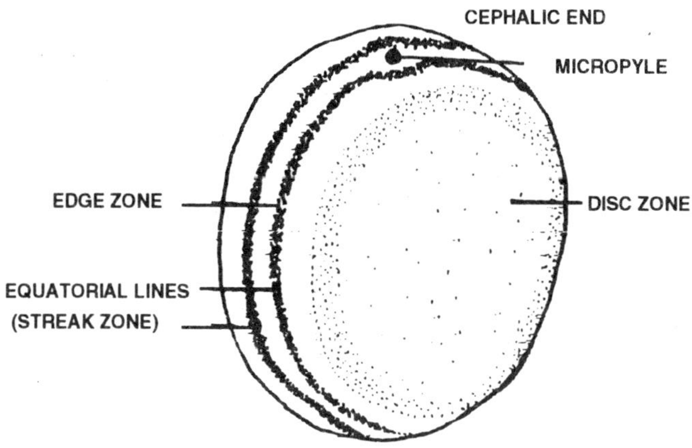

Fig. 3.1 : Structure of Tasar Egg

3.1.2. Larva

The larva is elongated, cylindrical being flattened on the ventral surface, robust, hairy and clothed with numerous tubercles (Fig. 3.2). Integument is soft and delicate in early stages but becomes tough and leathery in later stages. Hatchling's body colour is dull brownish-yellow with a black head. But the body colour turns green and the head brown with bright ornamentation. Yellow, blue and almond coloured larvae are also found rarely. The body of the larva can be precisely distinguished into three distinct regions : (i) head (ii) thorax and

(iii) abdomen. The latter two regions are composed of three and 10 segments respectively. The thorax and abdomen are similar in their general appearance. The caterpillar has secondary hairs which develop in the latter stadia and obscure the arrangement of the primaries.

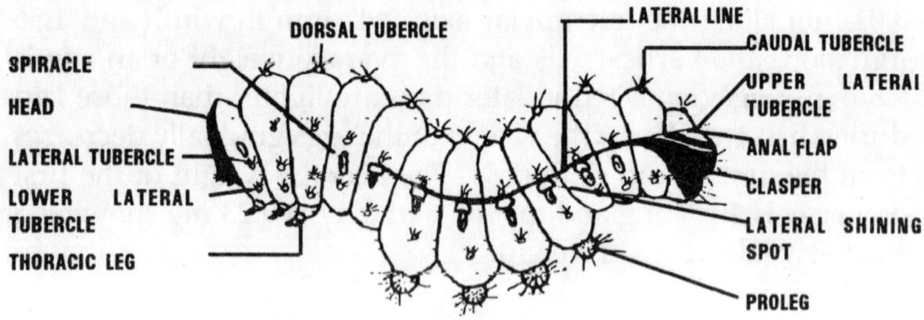

Fig. 3.2 : Structure of Tasar Larva

Head

The caterpillar has a well defined sclerotic and highly chitinised head with biting and chewing parts. The head being retracted into the prothorax, the dorsocaudal portion of the head capsule is not as heavily sclerotised or as deeply pigmented as the exposed portion. There is no sclerite between the antenna and mandibles. The head is hypognathus, prominent, ovoid, compact, free and readily mobile with a number of gnathal appendages above the oral opening. The most striking feature in the face of caterpillars is a deep inverted V shaped inflection consisting of a mid-cranial ridge proximally and two diverging arms distally which run obliquely to the neighbourhood of the anterior mandibular articulation. Externally these two arms form an inverted V shaped sulcus (Fs) which embraces the two sides of the triangular median sclerite the *frons* (Fig. 3.3). The anterior tentorial pits lie in that sulcus near the middle of its arms. The adfrontal sulcus is found to run along the inverted V shaped sulcus, thus marking the outer boundary of a narrow and paired sclerites the *adfrons* (ADF). Bordering the lower side of the frons is the membranous clypeus (CL). Suspended from the clypeus by means of an articular line, there is a single unpaired sclerite, the labrum (LM), forming the anterior most sclerite of the head capsule. In the posterior ventral region of the head

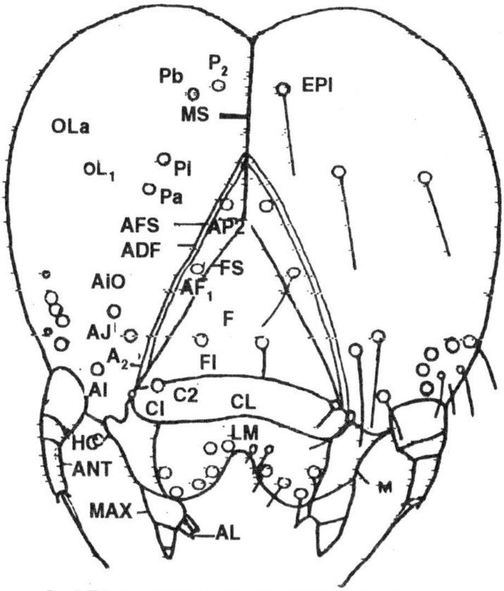

A. CEPHALIC VIEW OF THE HEAD

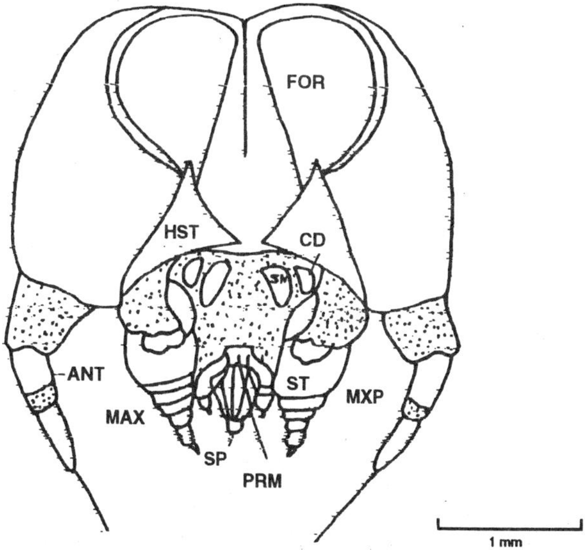

B. CAUDAL VIEW OF THE HEAD

Fig. 3.3 : Cephalic and Caudal View of Tasar Moth Head

(Fig. 3.3B) there is the mesal extension of the hypostomal areas of the cranium, thus forming the hypostomal lobes (HST). Each of the hypostomal lobes is separated from the corresponding post genal region by a distinct hypostomal suture which extends posteriorly to the root of the posterior arm of the tentorium concealed in the deep inflection at the base of the neck. The hypostomal lobes do not unite to form a hypostomal bridge unlike in the head of the adult Hymenoptera, Diptera and Heteroptera. The hypostome carries the maxillolabial complex (Fig.3.4,5) and separates its base from the neck.

Two kinds of setae and punctures are recognised in the cranium. The primary setae and punctures are present in the first instar larva. But secondary setae and punctures develop in the latter instars which obscure the arrangement of the primaries. Each epicranium of the first instar larva consists of three primary setae and six punctures (Fig. 3.3A). There are three anterior setae, the seta A_1 is situated near the base of the antenna, A_2 is slightly proximal to A_1 whereas A_3 is inclined towards the ocelli and lies slightly posterior to A_2. A_3 is the largest, but A_1 and A_2 are almost of equal length. The puncture Aa is situated in the anterior part of the head. There are two posterior punctures P_1 and P_2. P_2 is shorter than P_1 and situated proximally closer to the midcranial inflection. The punctures Pa and Pb are situated in the posterior region of the epicrania, but the latter is closer to P_2. One lateral seta L_1 and the puncture La lie litter posterior to it. The ocellar setae O_1 and O_2 are present near the 5th and 2nd ocelli respectively, while O_3 is in the horizontal axis of the first ocellus. The puncture Oa is situated between the first ocellus and the seta O_3. There are also three subocellar setae arranged disorderly. In the genal region, there is a single seta G_1 and adjacent to it is the puncture Ga.

Besides, there are two hypeal setae C_1 and C_2 situated near the extremity. C_2 is shorter than C_1 and situated towards the median line. Frons have only one seta F_1 and one puncture Fa lies lateromesal to it. The adfrontal carries two setae AF_1 and AF_2 and a puncture AFa. The seta AF_2 is situated near the posterior extremity of the sclerite while AF_1 is at the middle. The puncture AFa is situated nearer to AF_2 than to AF_1. Each epicranium has six ocelli arranged more or less in the form of a

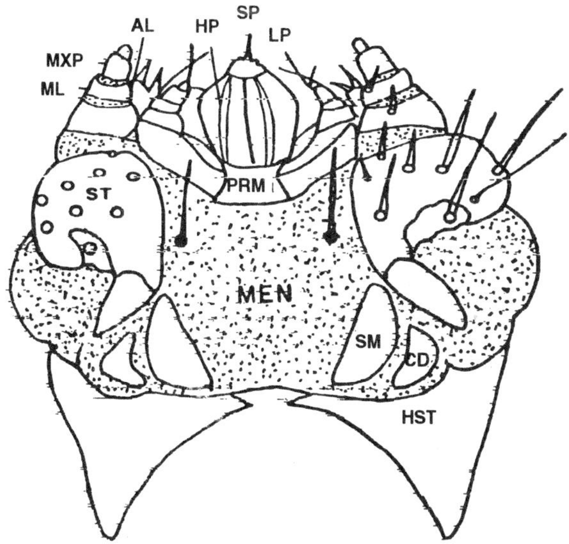

Fig. 3.4 : Maxillo - Labial Complex

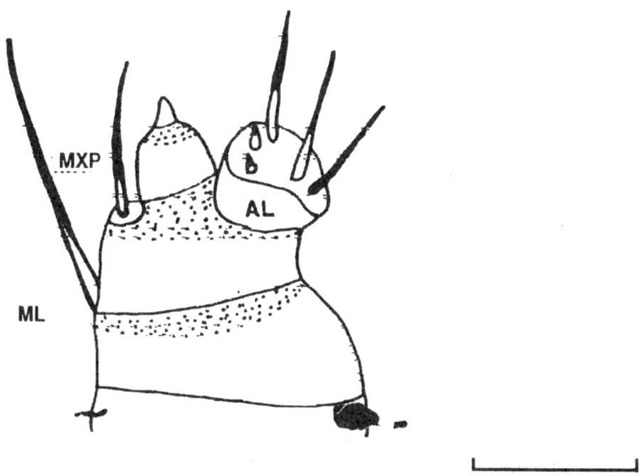

Fig. 3.5 : Maxillary Lobe (Inner View)

comma (Fig. 3.6A). They are brown to shiny black in colour and each in the form of hemispherical protuberant wart. The 2nd, 3rd and 4th ocelli are placed close together while the 6th one is located further away near the base of the antenna (ANT).

Each antenna is lodged in a pit of the head capsule known as *antennal socket* which is located at the base of the mandible (M) and extends nearly up to the ocellus. The antenna is short and three segmented (Fig. 3.6B). The first segment or the scape (SC) is broad, large and enclosed in a membrane. The second segment or the pedicel (PD) is shorter than the scape while the 3rd segment or flagellum (FL) is cylindrical in shape and larger and narrower than the pedicel. The scape is lighter in colour than the rest of the antenna. Two setae and three sensory structures (S) present at the tip of the flagellum (Fig. 3.6C). The proximal seta is short while the distal one is very long. One of the sensory structures is located near the long distal seta while the other two are mounted on a tubercle.

The mouth parts are typically of the biting and chewing type consisting of a labrum, a pair of mandibles and a maxillo-labial complex. The labrum is an unpaired, oblong, wider and a flat sclerite with little convexity (Fig. 3.6F, G) and is marked off from the clypeus by an articular line. The free end of the labrum is rounded and is provided with a deep median notch covering about three fourth of its depth. It has a row of four setae across the top of the notch (i and ii), two pairs of setae on the lateral edge (iii and iv) and two pairs of setae on the inner edge of the notch (v and vi). On the epipharynx (EX) there are two sets of marginal sensory papillae, each consisting of three sensillumbasiconica (SB) situated on either side of the median notch. The posterior boundary has a hingelike thickening the *tormae* (*T*) at each end.

The mandible is paired and each consisting of single dark brown solid piece of sclerotised cuticle situated on the lateral margins of the preoral cavity (Fig. 3.6D, E). Each mandible is articulated ventrally in the cranial margins of the head by a hypocondyle (HC) and dorsally through the epicondyle (EC) to which the apodemes of the abductor and adductor muscles are attached. Each mandible is found to bear five protuberances in the form of teeth, three of which are incisors (I) and two are

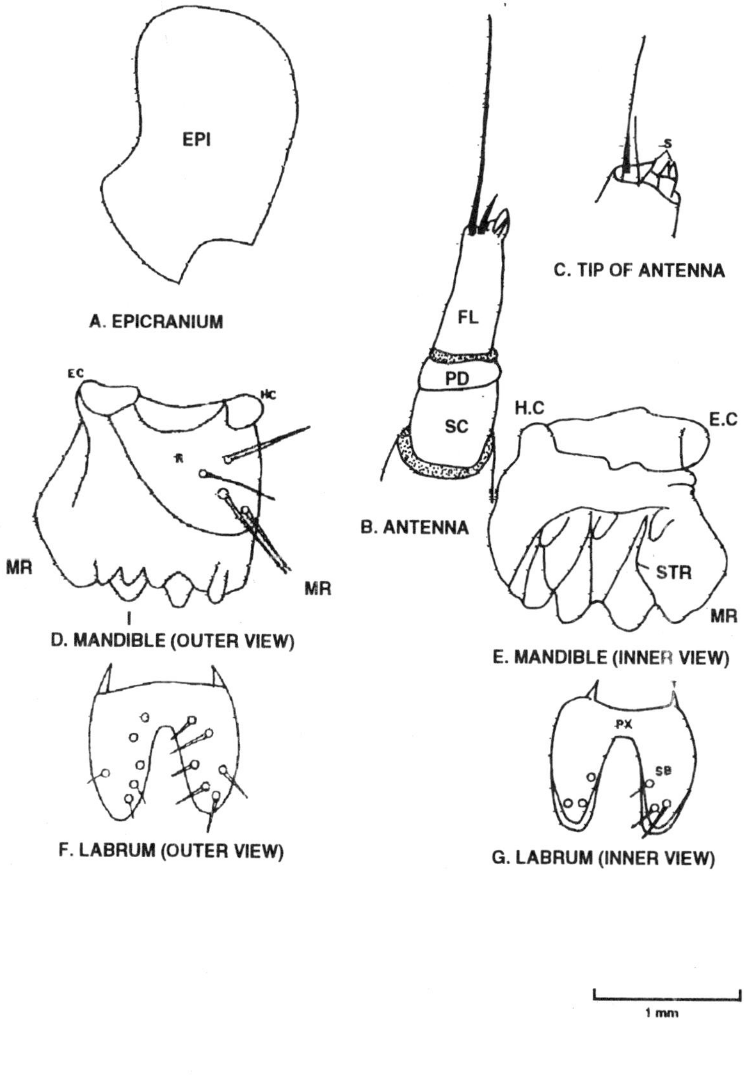

Fig. 3.6 : Cephalic Appendages of Tasar Larva

molar (MR). The upper portion of the mandible is lighter in colour than the toothed portion. Each mandible has a ridge near the outer margin having four white setae, three of which form more or less a triangle while the fourth one is placed towards the ridge. The inner surface of the mandible is concave with curved striations (STR) running from the denticulated ridges and finally reach the inner distal angle of the mandible.

The maxilla and the labium including the hypopharynx are united at the base to form a large composite maxillo-labial complex that projects behind the mouth and carries the spinneret at the tip (Fig. 3.4). These are supported basally on the hypostoma (HST). The maxillae (MAX) are presented by a pair of lateral lobelike structures which are distinctly marked from the median lobe representing the labium and hypopharynx. These are heavily chitinized and are glove shaped. Articulating with the hypostoma is the undivided cardo (CD). Mesial to each maxillary lobe (ML) is an internal ridge, lateral to which is situated the stipes (ST). The latter runs forward adjacent to cardo and then bends outwards. The stipes have nine white setae. The maxillary lobe (ML) is well developed and bisegmented. The proximal segment has a marginal seta while the distal one carries the bi-segmented maxillary palpi (MXP). The basal segment of maxillary palpi is large and short while the other is a small piece at the tip. Mesial to it is the accessory lobe (AL) bearing two sensory papillae and three white setae. The distal segment has only single white seta.

The median component of the maxillo-labial complex is composed of the labium and hypopharynx (HP). The labium is more or less a ring shaped structure composed of three sclerites, the submentum, mentum and prementum. The submentum (SM) is hemispherical, heavily chitinized and deeply pigmented plate at the base of the maxilla. The mentum (MEN) is much larger than the former but lightly chitinized with two pairs of setae. The longer pair is situated proximal to prementum and the other is the angular curve of the stipes. The prementum (PRM) is distally located and is represented by a free median lobe between the terminal lobes of the maxillae with the hypopharynx adnate upon its anterior surface. Distally the prementum carries the spinneret (SP). The latter is longer than wide and its base is surrounded by a narrow semicircular chitinized ring. Bi-segmented labial palpi (LP) are situated on either side of the spinneret. The basal segment of the palp is long and stout, while the distal one is a small piece at the tip and has two setae, one long and the other short. Secondary hairs develop on the maxillae but there is no such development on labrum, mandible and the labium.

Thorax and Abdomen

The segments of the thorax and the abdomen are very much alike especially in the early stages. These segments are more or less distinctly subdivided transversely into subsegments or annulets. The first thoracic and the last abdominal segments are the most modified structures among all the segments.

Thorax

The dorsum of the first thoracic segment is covered by the prothoracic shield. The shield is broad and divided into two equal halves by a middorsal longitudinal fissure. The prothorax has a pair of spiracles laterally near its hind margin. The meso and metathoracic segments are devoid of spiracles. Each of the three segments is provided with a pair of true legs. The legs are curved, four jointed and the basal joint is fleshy while the distal ones are leathery and horny. The terminal joint is armed with a small curved simple unguis or claw (CL). All the thoracic segments are armed with spiracles.

Abdomen

Abdomen is 10 segmented and each of the first eight abdominal segments is provided with a pair of lateral spiracles. Third segment to sixth and tenth segment are provided with a pair of prolegs (Fig. 3.7A, B) out of which first four pairs are the ventral prolegs or claspers. Each of the ventral prolegs has a fleshy conical projection known as *plants* (PL) which is more or less retractile and flat. A series of hooks or crochets (CR) are present around the edge of the plants. The hooks are 84 in number in the ventral prolegs of the mature caterpillar and are arranged biordinally in meso series (Fig. 3.9C). The anal pair of proleg or clasper is directed both downwards and backwards and supplied with numerous hooks. The hooks are also arranged

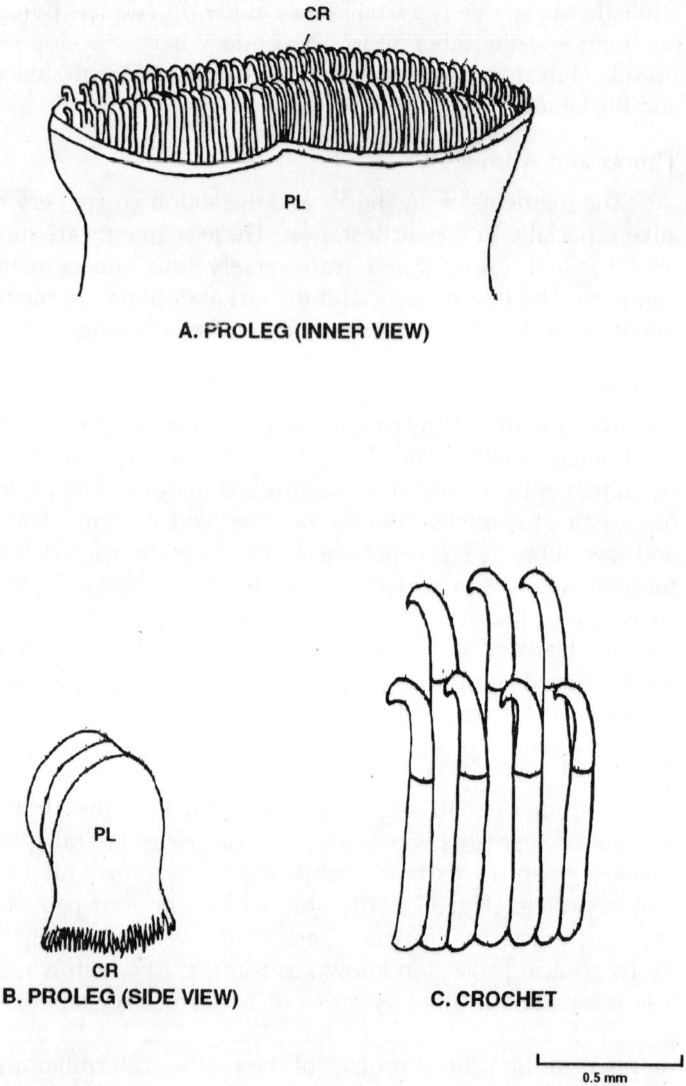

Fig. 3.7 : Structure of Proleg and Crochet of Tasar Caterpillar

in a similar fashion as in ventral prolegs. The 10th abdominal segment forms the suranal plate. The paranal lobes are fleshy and appear as if projecting backward from the base of the anal prolegs. Each of the first eight abdominal segments has a pair of distinct spiracles. These spiracles have thickened lips and are oval in shape with a raised margin. The lips are fringed with branched processes. The spiracles of the prothorax and the 8th abdominal segment are larger than those of the other segments, but the shape of the prothoracic spiracle is distinctly different from the rest of the spiracles.

Tubercles

Four types of tubercles are noticed in the body : (i) dorsal tubercle (ii) super lateral tubercle or supraspiral tubercle (iii) lateral tubercle or subspiral tubercle and (iv) lower lateral tubercle. The prothorax is provided with a pair of dorsal lateral and lower lateral tubercles on both sides of the body (Fig. 3.8A). The meso and metathorax have four pairs of tubercles (Fig. 3.8 B) such as dorsal, upper lateral, lateral and lower lateral tubercles on both sides of the body. First segment to 7th segment of the abdomen carry three pairs of tubercles each except the lower lateral one (Fig. 3.8D). In the 8th abdominal segment (Fig. 3.8C) all the three types of tubercles are distributed but the dorsal tubercle is fused at the base to form a caudal tubercle. The upper lateral tubercle is pushed back a little and thereby is not in like with those of the preceding segments. The 9th abdominal segment has only two pairs of tubercles (Fig. 3.8E) as dorsal and lateral tubercles; the others being atrophied. In the 10th segment, the dorsal and lateral tubercles are united to appear like a single entity and the other tubercles atrophy (Fig. 3.8F).

Hairs and Setae

The hairs are small, white and irregularly distributed. The setae are of two types such as body setae and tubercular setae. The body setae are long, pointed and white at the base of the dorsal tubercles from first to 8th abdominal segment. The tubercular setae are also long, pointed but black in colour which are present at the tip of the tubercles. The prothoracic hood is provided with setae just like tubercular setae at the corresponding position of the dorsal tubercle. The number and arrangement of

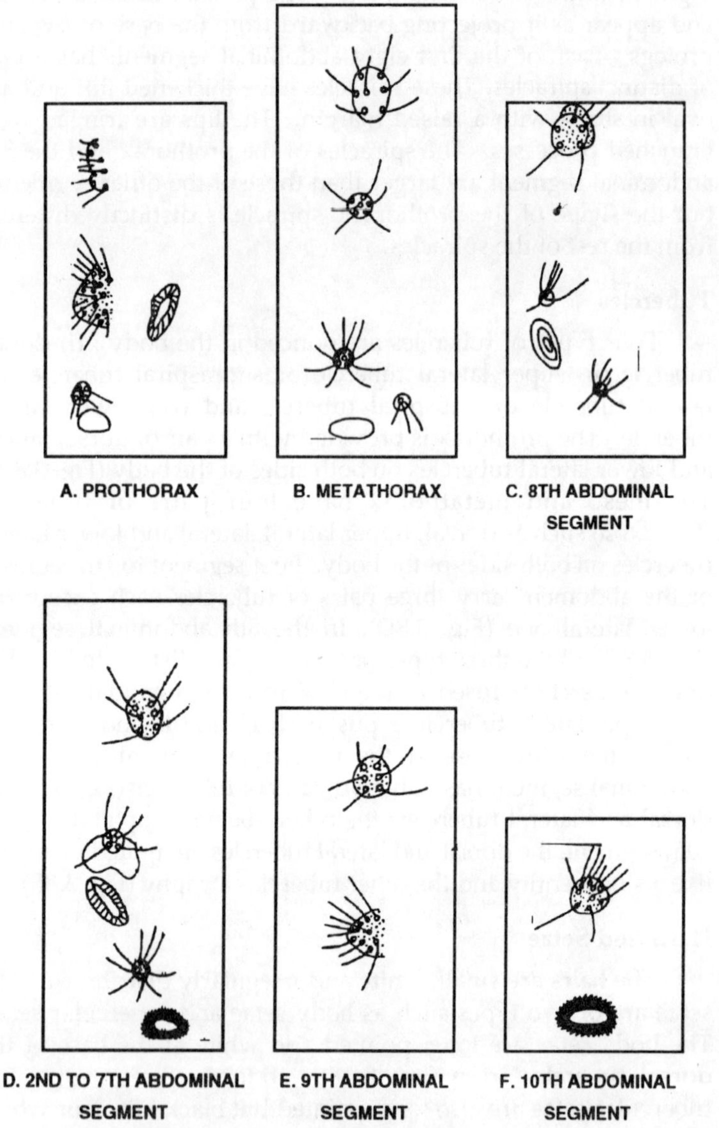

Fig. 3.8 : Chaetotaxy of Tasar Caterpillar

the setae vary according to the type of tubercle, body segment and age of the larvae. The total number of tubercular setae remain constant until the 3rd instar and diminishes thereafter.

Prothoracic Hood

In hatchings, prothoracic hood dorsally carries an oval black spot enclosing two semilunar white markings. These markings disappear during first instar. In early second instar, the black spot splits to an M-shaped structure and subsequently becomes V-shaped in appearance with two dot markings. These marks are absent in the 3rd instar and again reappears as two semilunar red markings corresponding to the epicranial plates in 4th and 5th instars (Fig. 3.9).

Dorsal Line, Lateral Line and Ventral Line

A black middorsal line is present in the hatching and in the first instar extending from first to seventh abdominal segment and a dumpbell-shaped black mark on 8th segment. The black line disappears during late second instar. In the hatching, black perpendicular double lines are present in the abdominal segments from 1st to 8th. These lines become V-shaped during the first instar and disappear in the 2nd instar. During 3rd instar a yellow line develops laterally from 2nd to 10th abdominal segment and in the 4th and 5th instar another brown line also develops over it. In all the instars, that is from 1st to 5th a red mid ventral line is present which extends throughout the abdomen.

Anal Flap

The anal flap is a triangular black mark during hatching and first instar which turns brownish and becomes V-shaped in appearance from the 2nd instar onwards. The triangular mark on each of the claspers is black in first and second instar and brown in other instars (Fig. 3.9).

Shining Spots

In the 3rd instar from 2nd to 7th abdominal segments, silvery white lateral shining spots in the oval or triangular form appear at the foot of the upper lateral supraspiracular tubercle. Six regular and 38 irregular patterns are noticed. Larvae without

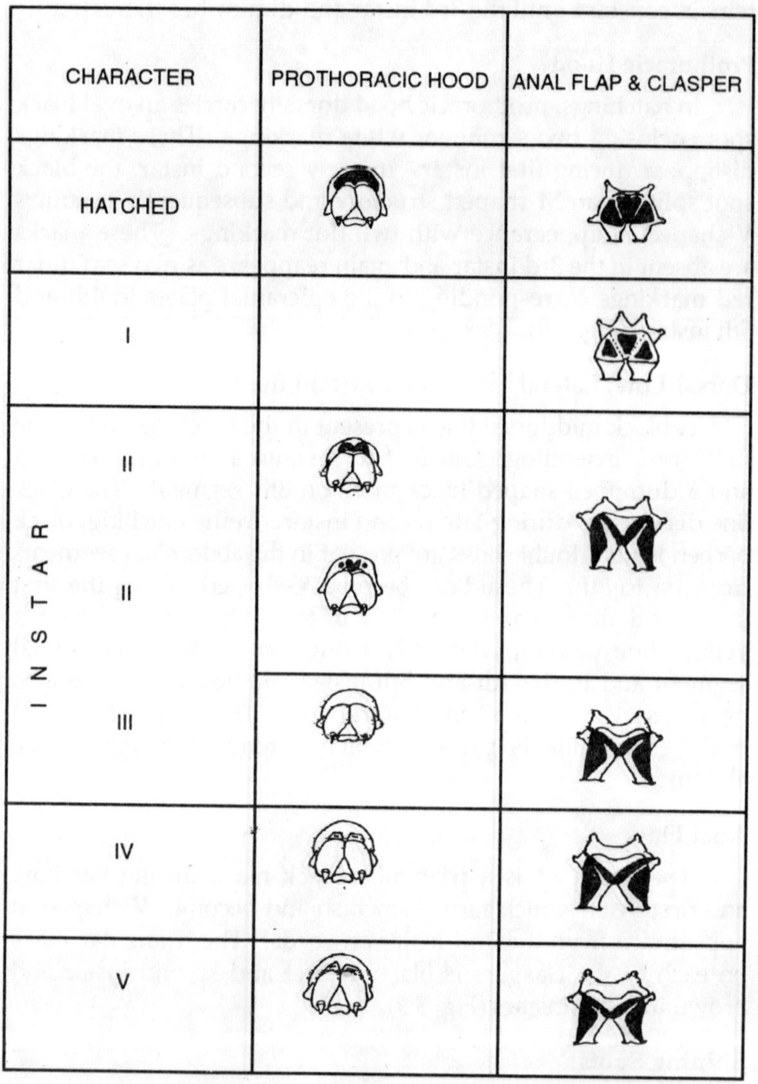

Fig. 3.9 : Prothoracic Hood, Anal Flap and Clasper of Instars

any shining spots are also seen in natural tasar biodiversity. Other shining spots are developed at the base of the dorsal tubercle. These are brick red in appearance in green and yellow larvae and white in blue and almond coloured larvae.

3.1.2.i. Just Hatched Larva

The body of the newly hatched larva is brownish yellow in colour. Head is shining black or deep brown, comparatively large bearing 13 primary setae and six punctures. The prothorax is the largest part followed by metathorax and mesothorax. Mesothorax is very much compressed in between pro and metathoracic segments. The shieldlike pronotum has a prominent black patch extending transversely. The dorsal tubercle is yellowish white in colour bearing four transparent white hairs. The lateral tubercle, having 20 long white hairs arranged in a whorl, is larger than the dorsal and in fact, the largest of all the body tubercles (Fig. 3.10A). It has a lustrous yellow base with the apex yellow and brownish tinge. The lower lateral tubercle at the leg base is of the body colour. It has five white hairs which are comparatively lighter in colour. Of the tubercles present on the rest of the thoracic segments, the dorsal one is the largest. In the second thoracic segment it is brownish yellow in colour with eight dark brown setae. The upper lateral tubercle or supraspiracular is yellow in colour with six light brown setae. The lateral or subspiracular tubercle is yellow with nine white setae. The lower lateral tubercle is of the body colour and has three white setae. The dorsal tubercular wart of the metathorax (Fig. 3.10B) is the largest of all the body dorsal tubercles. It is prominently black with light black setae. The upper lateral tubercle is yellow with six black whereas the lateral tubercle is yellow with 10 white setae. The lower lateral tubercle is that of the body colour with three white setae. In the abdominal segments from 1 - 7,.the dorsal tubercle is yellow with brownish tinge at the apex and has six brown setae. The upper lateral tubercle is also yellow with five brown setae. The lateral tubercle is yellow having nine white setae.

The caudal horn is black in appearance with 10 black setae (Fig. 3.10C). In the 9th abdominal segment, the upper lateral or supraspiracular wart is larger than the dorsal tubercle. Dorsal

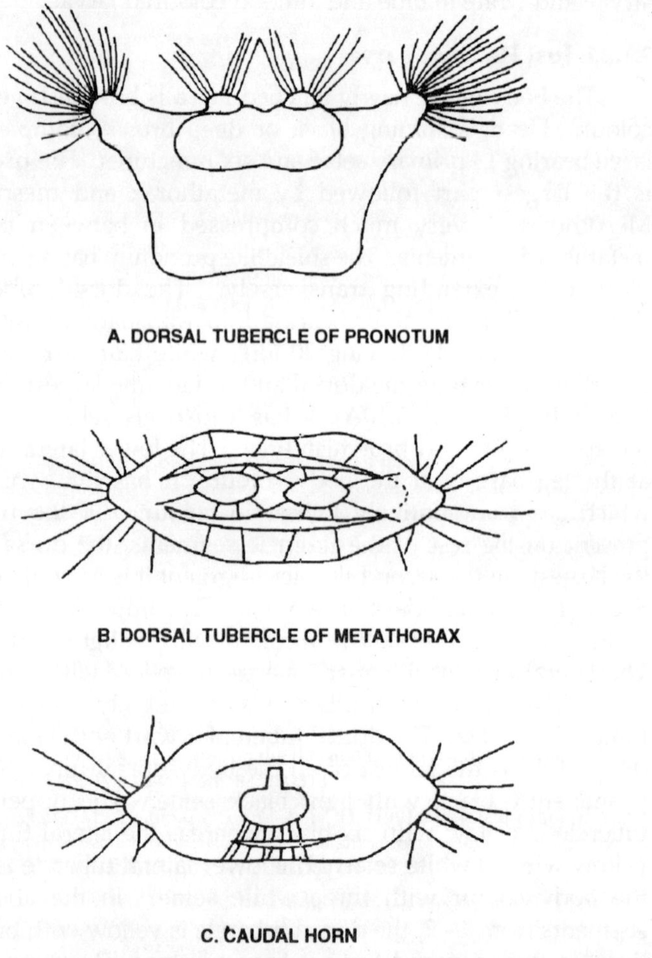

Fig. 3.10 : Tubercles of Tasar Hatchling

tubercle is yellow with six light brown setae, while the supraspiracular tubercle is yellow with 10 white setae. The dorsal tubercle of the 10th segment is light yellow with 10 white setae. The dorsum of the 10th segment has a black shieldlike patch. Two chocolate coloured bands running across the body of the larva on the dorsal surface on either side of the dorsal and upper lateral tubercle darken from the 3rd thoracic segment onwards to form six distinct black patches in each segment. Two of these patches are located in between the dorsal tubercles and rest four are in between upper lateral and lateral tubercle of either side. The spiracles in the body are normally black except the prothoracic one which is comparatively lighter in colour. The thoracic legs are black with 5 - 6 small white setae while the prolegs have black horizontal bars just above the extremities. The colour of the larva gradually becomes greenish as it proceeds towards the second stadium. The number of setae in different tubercles remain constant upto the end of 4th moult.

3.1.2.ii. Second stadium

In the second stadium, the body colour becomes lemon-yellow. Head is burnt amber and the dorsocaudal portion is lighter than the anterior exposed portion. The midcranial inflection and the adfrontal sulcus become distinct. The frons is brown, palp is light grey and the antenna is yellow-ochre in colour. A number of setae develop irregularly mostly in the anterior portion of the epicrania. The setae on the head are all black. The single black shieldlike patch on the pronotum splits into either 2 or 4 distinct bits. The occurrence of these two groups of patches is observed in larvae from a single laying. When splitted into four bits, they are arranged two in a group located at the base of dorsal tubercle of the four bits, the middle two are bigger and more or less triangular in shape and bend towards each other, whereas the distal two are small and irregular in shape. When splitted into two bits, they are irregular in shape, much reduced in size and located at the base of the dorsal tubercle. The dorsal tubercular wart is yellow with black setae extending over the epicranium (Fig. 3.11A). The lateral tubercular wart is greenish yellow with distinct orange-red apex bearing black setae arranged in a whorl. The tubercle at the leg base is lemon-yellow in colour with black setae. The tubercles

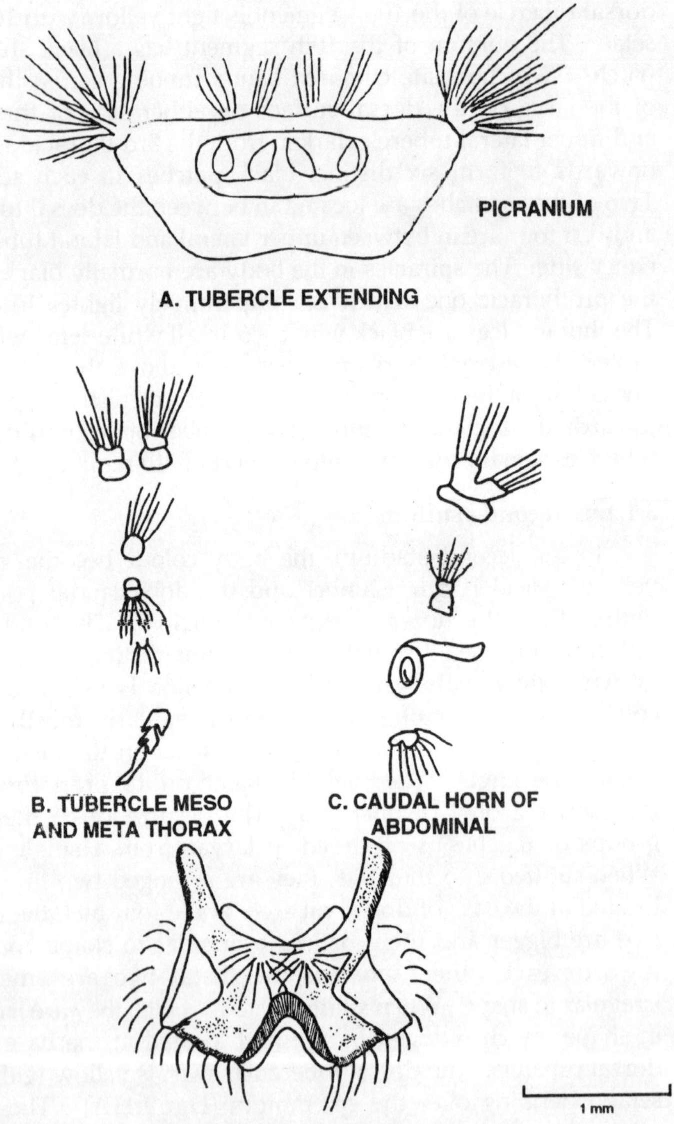

Fig. 3.11 : Tubercle of Second Instar Tasar Larva

in the meso and metathorax are more or less similar in colour (Fig. 3.11B). The dorsal tubercle has distinct orange base and black apex with black setae. The supraspiracular or upper lateral tubercle is basally orange in colour with the apex black bearing black setae. The subspiracular or lateral tubercle is basally greenish-yellow and apically orange in colour with black setae. The lower lateral tubercle is that of the body colour with white setae.

The abdominal segments from 1-7 have similar tubercular warts arranged transversely. The dorsal tubercle has orange coloured base and black apex with black setae. Supraspiracular tubercle is similar in colour as the dorsal one. The subspiracular tubercle is orange in colour with a narrow yellow base bearing brown setae. The caudal tubercle of the 8th abdominal segment (Fig. 3.11C) is black with orange coloured apex bearing black setae. Supraspiracular wart is orange with the apex having black tinge. The subspiracular or lateral tubercle is orange-red in colour without black apex and is having light brown setae. The upper lateral tubercle of the 9th segment is orange-red in colour with black setae (Fig. 3.11D). The single triangular shaped black shield on the suranal plate becomes divided into two black patches converging posteriorly. The dorsal tubercle is located at the posterior extremity of the suranal plate and is dull brownish in colour with black setae set in a whorl. The posterior part of the suranal plate is fringed with numerous light brown and white setae. A number of secondary white setae begin to develop and are distributed irregularly. The spiracle is black in colour with chocolate peretreme. The legs are black with 10-12 white setae. The prolegs have black parallel markings.

3.1.2.iii. 3rd instar

The body of the larva during 3rd instar turns leafy-green. The head is chocolate in colour with distinct midcranial inflection and adfrontal sulcus, frons are sienna and the palpi are grey in colour. A large number of secondary setae develop in the head thus, obscuring the arrangement of the primaries. The dorsal tubercle of the prothorax is slightly yellowish green bearing eight black setae in two groups of four each (Fig. 3.12A). The lateral tubercular wart is prominent and has orange coloured base and bluish - grey apex with black setae arranged in a whorl. The

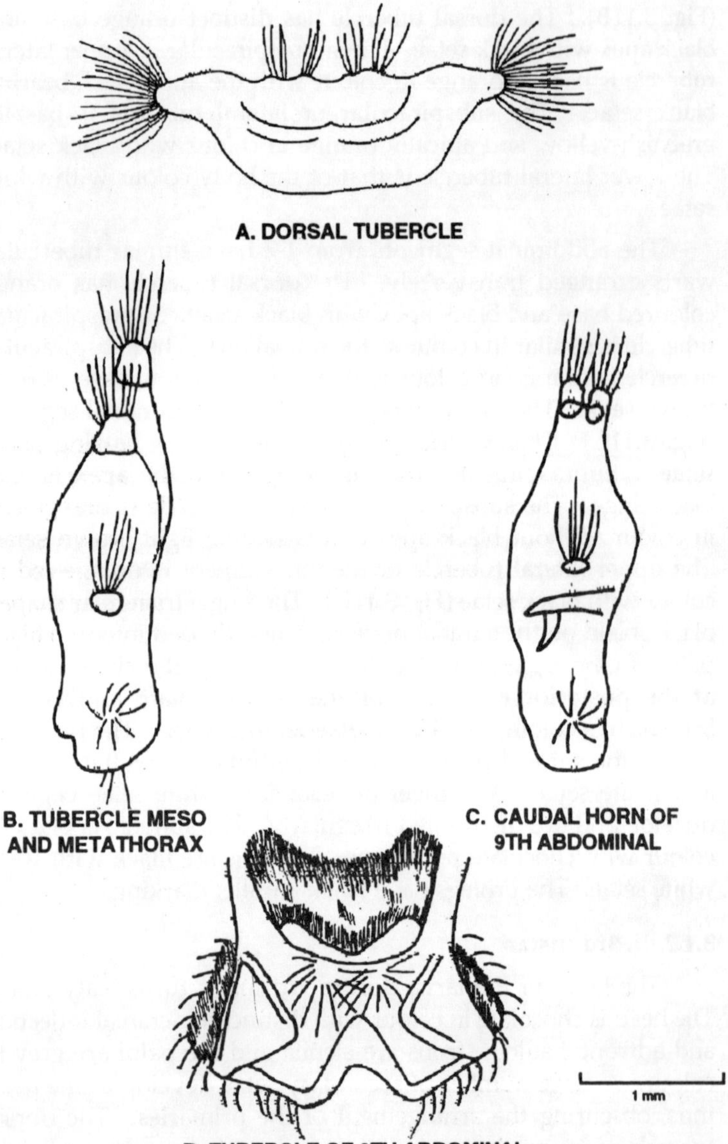

Fig. 3.12 : Tubercle of Third Instar Tasar Larva

lower lateral tubercle is also of the body colour with black setae set in an irregular pattern. The tubercles of the meso and metathorax (Fig. 3.12B) are mostly similar except that the dorsal tubercle of the metathorax is the largest of the dorsal tubercles present in the thoracic segments. The dorsal tubercle is orange-yellow in colour bearing eight black setae—six marginal and two central. The tubercle has orange-red colour base and violet apex bearing six black setae in three groups of two each. The subspiracular or lateral tubercle is bluish in colour with nine black setae in three groups of three each. Two lower lateral tubercle is of the body colour with three white setae set irregularly. The abdominal segments from 1-7 have similar tubercular warts. The dorsal tubercle is light orange coloured at the base with black apex. The upper lateral tubercle has light orange coloured base with violet tip. Subspiracular or lateral tubercle is bluish in colour with nine brown setae in three groups of three each, one being at the centre. The caudal tubercle is basally orange-yellow and apically black with 10 setae in two groups of five each (Fig. 3.12C). The supraspiracular and the subspiracular tubercles are similar in colour and setal number. Their position is as in the abdominal segments 1-7. The dorsal tubercle of the 9th segment is orange-yellow in colour with dark brown apex bearing six brown setae (Fig. 3.12D). The upper lateral tubercle is larger than the dorsal tubercle and has 10 brown setae set in a whorl. The dorsal tubercle in the 10th segment is light green in colour with 10 brown setae arranged in a whorl. The suranal plate is fringed with numerous light brown setae posteriorly.

More number of secondary white setae develop on the body segments. The metallic spots develop enveloping the lower part of the upper lateral tubercle and are distributed in the 2nd, 3rd and 4th abdominal segments but are not very conspicuous. An inconspicuous brownish-yellow band runs from the 2nd to the 9th abdominal segment just above the spiracles. The anal angle is marked with a chocolate triangle. Spiracles are chocolate in colour except in the 8th abdominal segment. The peretreme is light brown. The middle two segments of the legs are dark brown but the first and the terminal joints are brownish-yellow. The number of setae in legs are more than that in the 2nd instar (Fig. 3.13). The prolegs are of the body colour except the distal

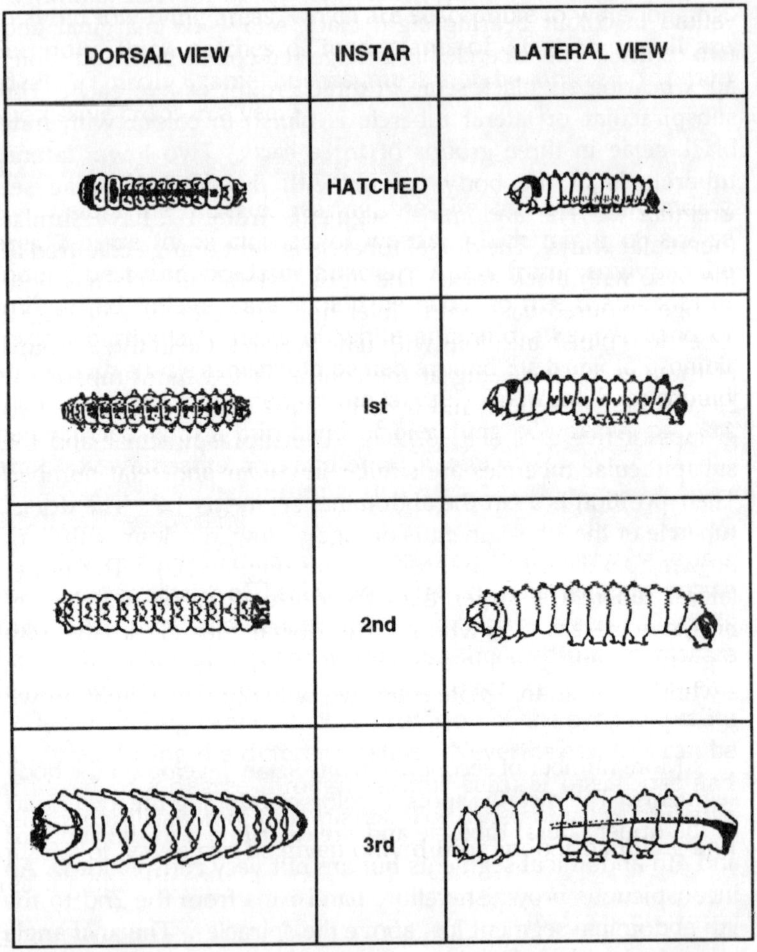

Fig. 3.13 : Developmental Morphology of Tasar Larva

portion bearing crochets, which is light brown. The prolegs are provided with large number of black setae.

3.1.2. iv. Fourth Stadium

After the 3rd moult the larvae turn to deep green having burnt-sienna head. The secondary setae that develop on the head during the 3rd instar become longer. The dorsal tubercle on the prothorax is greenish-yellow with black setae. The number of setae and their arrangement in all the tubercles are same as in previous instars. At the base of dorsal tubercles there are a few irregular patches, orange-red in colour. The lateral tubercle is basally green and apically greenish-yellow and much larger than the dorsal tubercle. It has 20 black setae arranged in a whorl. The lower lateral tubercle is of the body colour with black setae.

The dorsal tubercle of meso and metathorax is lustrous orange-red with the apex black bearing black setae. The supraspiracular tubercle is violet with pinkish base bearing six black setae. The subspiracular tubercle is bluish in colour with nine black setae. The lower lateral tubercle is of dull bluish colour having three or four hairs. The dorsal tubercle of the abdominal segments from 1 - 7 is lustrous orange-red with black apex bearing six black setae. The supraspiracular tubercle is pink at the base with violet apex bearing five black setae. The subspiracular tubercle is violet in colour with nine black setae. The caudal tubercle is of the body colour and apically slightly black having black setae. The dorsal tubercle of the 9th abdominal segment is light bluish having six brown setae. The supraspiracular tubercle is larger than dorsal tubercle and has 10 brown setae. The dorsal tubercle of the suranal plate is bordered with light green and burnt-sienna and has a tuft of black hairs.

The number of secondary hairs on the body is much more than in the 3rd instar. The brownish-yellow lateral band becomes more conspicuous and the brown anal triangle projects into the lateral line across 8th and 9th abdominal segments. The metallic spots at the base of the tubercle become more distinct. The spiracles are bold orange outlined with black. The legs are chocolate and the prolegs are green in appearance.

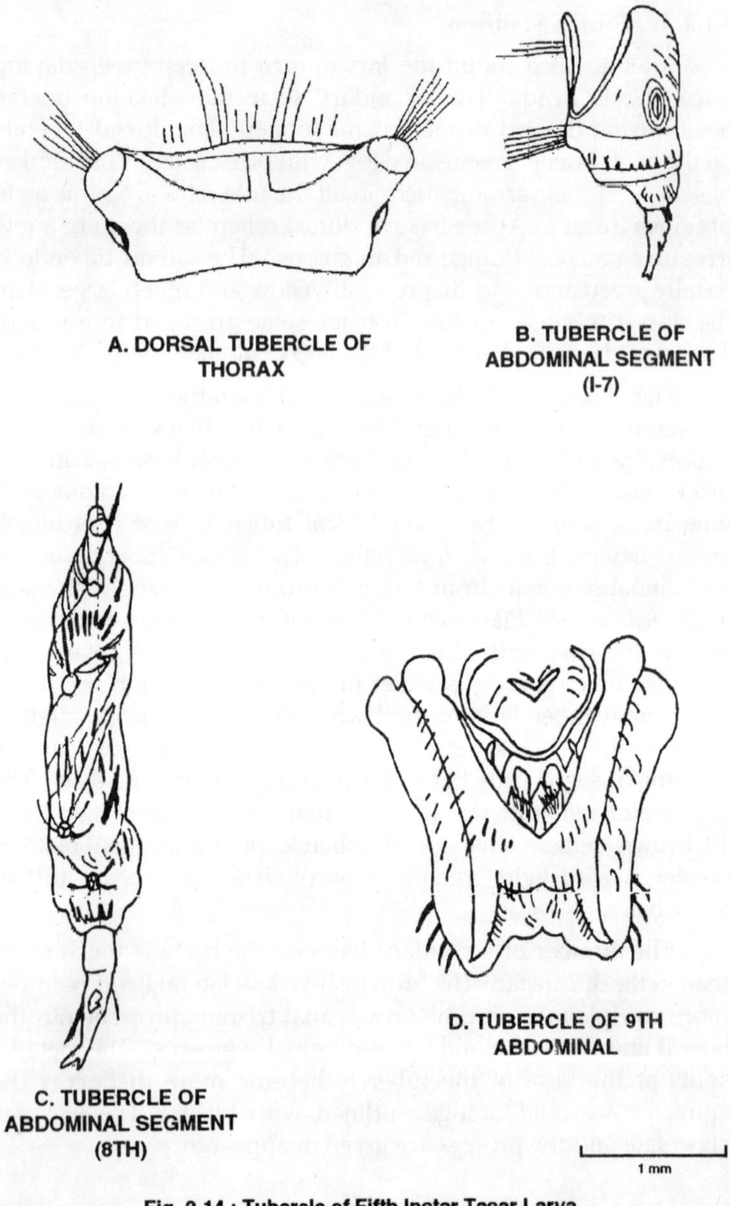

Fig. 3.14 : Tubercle of Fifth Instar Tasar Larva

3.1.2 v. Fifth Instar

After the fourth moult, larvae of two different colours are distinguished. A large number of larvae are apple-green while a few are hooker's-green or yellowish in colour. Head is chestnut with conspicuous yellowish-brown labrum and appendages. The colour of the tubercles are as in the 4th instar, but at the base of dorsal tubercles of prothorax, an orange-red band develops. The apex is, however, black (Fig. 3.14A). The setae on the tubercles are longer than before but reduced in number. In the abdominal segments from 1-7 (Fig. 3.14B) there are three sub-primary setae on either side of the mid dorsal longitudinal line in each segment. They are white in colour and almost as long as the primary setae. In the 9th (Fig. 3.14D) and 10th segment the sub-primary setae are absent. In the thoracic segment sub-primary setae are absent but the secondary setae are numerous and irregularly arranged. The distribution of metallic spots in the 5th stage of larva is not uniform. Two pairs of metallic spots occur most frequently followed by six pairs. Very rarely the larvae have been observed to possess four and five pairs of such spots.

On an average the larvae measures 0.71 cm in length, 0.28 cm in girth and 0.08 gms in weight in the first instar. But in 5th instar it measures 11.30 cm in length, 1.98 cm in girth and 25.18 gms in weight.

3.1.3. Pupa

The pupa is the third stage in the life-cycle and is oval in shape. General colour of the pupa is burnt-sienna and the whole surface is finely wrinkled. The body of the pupa is divisible into head, three thoracic and 10 abdominal segments. Ventral organs are burnt-sienna in colour and the abdominal incisions are yellowish-burnt-sienna. The head is narrower than the thorax, which again is narrower than the front abdominal segments. The abdominal segments increase in size from first to fourth segment and then gradually decrease up to the last segment. The dorsum possesses patches of brownish bristles.

Head

The vertex forms the dorsal area of the head behind the epicranial suture and anterior to it are present the frons and clypeus, frontoclypeal suture being distinct. The labrum is distinct but the clypeo-labral suture is not well defined. The glazed eye is very inconspicuous. The antenna cases are nearly as well developed in the female as in the male pupa with the axis of flagellum very prominent while in the imagines the antennae are only very alightly pectinated in female and only occupy a small portion of the head. Maxillary palpi are indicated as two short pieces lateral to eyes and they do not exceed 1/3rd the length of the wings. Antenna is shorter than the wing and measures 1.6 cm in length in both sexes.

Thorax

Three thoracic segments are distinct on the dorsum but ventrally they are concealed by the appendages. The prothorax is not well developed and is placed anteriorly. A fine smooth mid dorsal line runs down the centre of the thorax. Pronotum is narrow and rough. Mesonotum is the largest whereas the metanotum is ill developed. Wings are extended beyond the ventral half of the 4th abdominal segment and meet in the mid line. Forewing is provided with a thick costal margin, the hind pair appears as a narrow strip along the dorsal margin of the latter. Prothoracic spiracle is present in the suture separating the pro and mesothorax. The 1st and 2nd pair of legs are extended beyond the antenna. The thoracic segments are covered by numerous minute brown coloured bristles. One fourth of the second pair of legs and a portion of the 1st pair are covered by the wings.

Abdomen

First four abdominal segments are free dorsally but covered on the ventral surface by the integument of wing. Dorsally scars denoting the positions of the larval tubercles are visible on the abdominal segments. The spiracles are placed laterally and are conspicuous on the 2nd and 7th abdominal segments and aborted on the 8th. Each consists of a narrow split set in a prominent black rim. The spiracles of the first abdominal segment are covered by wings. The movable incisions are placed between 4 - 5, 5 - 6 and 6 - 7. Ventrally the impression of the prolegs is left on the 5th and 6th abdominal segment. Dorsal surface of 5th, 6th and 7th segment bears ridges. These lie in the anterior half of each segment between the lateral spiracles of the two sides. The genital aperture of the male pupa (Fig. 3.15A)

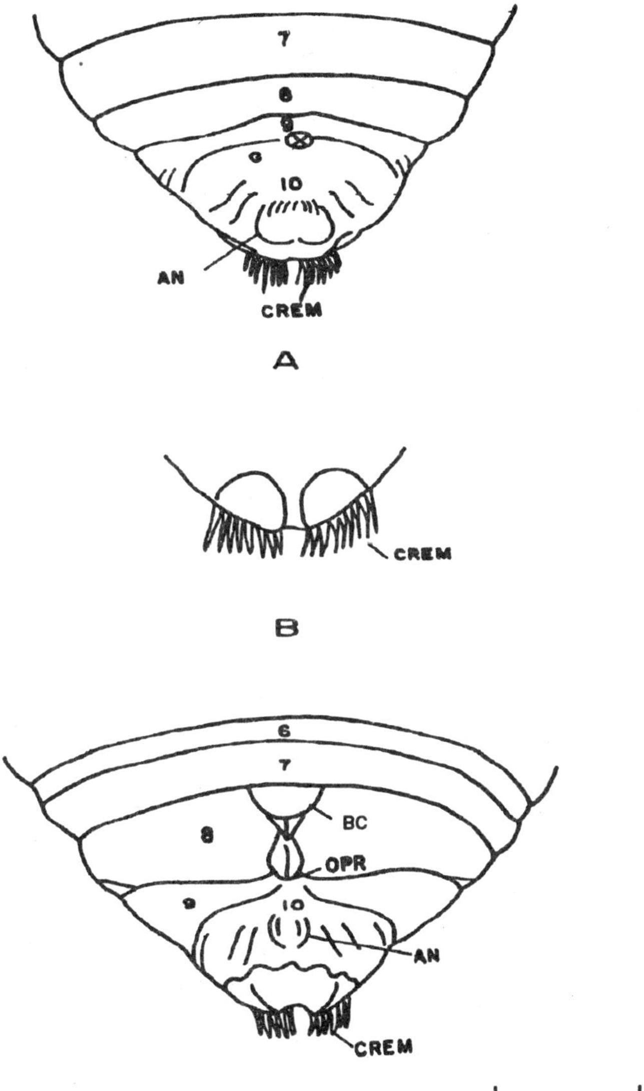

Fig. 3.15 : Abdominal Segment of Pupa of Tasar Moth

is situated on the ventral side of the 9th sternum. It is in the form of a fine slit lying between two raised projections. In the female the genital organs are clearly visible on the 8th and 9th abdominal segments (Fig. 3.15C). The anterior *bursa copulatrix* occupies the entire breadth of the 8th abdominal segment, its margin being very prominent. This posterior opening of the oviduct (OPR) similarly occupies the entire breadth of the 9th abdominal segment; its margin being not so distinctly marked as that of the anterior opening. The anus is placed on an oval convex area at the caudal margin of the 10th segment which forms the *cremaster* (Fig. 3.15B).

3.1.4. Cocoon

In sericigenous fauna, that is eri, muga, mulberry and tasar, the generalised cocoon structure with all the compliments is successfully found in *Antheraea mylitta* Drury and *Antheraea paphia* L. In *Antheraea pernyi* and *Antheraea roylei* peduncle is devoid of the loop. In *Philosamia ricini* H. and *Antheraea assamensis* W. the stalk is ill developed. In mulberry, eri and muga cocoons the peduncle is completely absent. Cocoons are master craft of the tasar insect prepared quite uniformly. These are oval shaped, closed, single shelled and light structures resembling small fruits with peduncle. Structurally cocoon has two distinct portions : (i) dark brown peduncle with a ring at the distal end and (ii) cocoon proper (Shell). Cocoons according to the colour pattern are of many types such as (i) black (ii) grey (iii) yellow (iv) greyish black (v) yellowish black (vi) cream and (vii) creamy white. Tasar cocoon is the protective covering spun by mature larva (5th instar) for the chrysalis and is also economically very valuable. Unlike eri, muga and mulberry, tasar cocoons are spun in nature and have unique distinction of a shell and a peduncle. Taking pupa, shell and peduncle into account, various types of cocoons are noticed in tasar crop.

As a remarkable portion of the silk insect, cocoon attracts the attention of scientists in the laboratory and field, a rearer in the field and grainage house, a purchaser in the market and women above all in the society. But how does silk insect prepare a cocoon is not known, but is believed to be an instinctive behavior. To start cocooning is a craftsman spirit for which silk insect may be called as *engineer or master craft insect of the natural fibre.* Tasar larvae are voracious eater of green leaves of the host plants which end at the last stage of larva or at maturity. End part of the feeding process is signalled by evaluation of the gut contents or passing of last excreta as a green semisolid mass followed by a colourless slimy substance. In the meantime, the larva selects the suitable site of the host plant particularly twigs having minimum three leaves very close to each other. Choosing the suitable portion of the plant, the last stage larva spends its final larval activities and becomes very much inert, dormant and motionless. Very often a cocooning zone is explored after about 10-25 minutes of rest and the larva may continue to search for one to $1\frac{1}{2}$ hours. Then the larva starts and completes its cocoon in four different phases such as (i) nest formation (ii) ring and peduncle formation (iii) cocoon proper (Shell) preparation and (iv) cementing.

(i) Nest Formation

Before the larva starts cocooning, it locates a part of the twig with three or more healthy leaves to provide the larva a support to build up the cocoon. The larva first links the leaves to form a small nest or *hammock* with an opening at the top by irregularly throwing out silk secretion from the silk gland. This liquid type of secretion hardens by coming in contact with air which appear to be thread-like structure. It continues to secret its silk haphazardly to the leaves till the leaves come closer to each other. For this process of nest formation, the larva takes about $5\frac{1}{2}$ to 6 hours. On bringing the 5th instar larvae from their original host plant to secondary host plant, they can also prepare hammock but the quality of cocoon becomes very thin.

(ii) Ring and Peduncle Formation

After completing the nest, the larva comes out of the hammock and with its mandibles straps bark 2-3 mm wide around the twig in the form of a ring within half an hour. The space within the ring is then filled up with silk. Since the bark is soft and rest part of the twig (Conducting tissue) is comparatively hard and tough, the larva debarks the twig and strengthens the ring space by filling up silk. Simultaneously, the peduncle formation is initiated by connecting the ring to the loop of the hammock with the help of few strands of silk. Gradually it is completed by further adding of silk strands.

During this mechanism the head oscillates to and fro and subsequently changes to semicircular and linear motions during the formation of ring and peduncle. These events take around 10 hours to be completed.

(iii) Cocoon Formation

After hammock, ring and peduncles being ready, cocoon proper is started. This is entirely a process of spinning just as a machine does from cotton. Usually larvae like to spin during day light hours. Spinning is made in a regular and systematic way during when the head movement of the larva forms a horizontal figure of English number eight (8). This systematic movement is carried out by the larva within the limited space of the initial cocoon and this movement continues till the entire body of the larva is invisible. The larva needs a day and night time (almost 24 hours) to envelop its body by its own silk secretion. During this stage although the larva is invisible from outside still then, it spins for another 2-3 days. So in course of time the cocoon becomes thick and hard. Moreover, the larva stops its continuity of being disturbed and the disturbed larvae produce the cocoons of low weight indicating less quantum of silk secretion.

(iv) Cementing

After cocoon is ready, the larva takes its position to take rest properly inside the cocoon. It keeps its head towards the peduncle end. Then it slowly discharges a small quantity of whitish-fluid or milky-yellow substance for 2-3 hours. Upon drying, the discharged fluid cement the cocoon which results in the colouration of the cocoon. In the wild condition the discharged fluid in excess is observed to be left on the soil of the host plants. This secretion probably has two functions : (i) protects the cocoon from rain water entry and (ii) hardens the cocoon thereby providing a safe space for the pupa. This is followed by pupation when the larval morphology gets changed with several other physiological changes.

3.1.4.i. Biometry of Cocoons

Although all the cocoons in the field or garlanding stage are similar in appearance, but there is difference between male and female cocoons in length, width and volume. The average dimension of male and female cocoons is 5.1 cm x 2.9 cm and 5.5 cm x 3.3 cm respectively. The peduncle length is also comparatively larger in size in female (5.7 cm in female and 4.0 cm in male). The average volume of the male cocoon is 28.6 cc whereas that of female is 29.5 cc. Thus, in all the respect, male cocoons are comparatively smaller in length, width and volume. Modal cocoons cultivated in the highest altitude (601-1,000 m ASL) in rainy season weigh the most (13.58 gm in male and 13.57 gm in female). Between Bogei and Nalia, the weight of the Nalia cocoons are more than Bogei both in rainy and autumn season.

3.1.4.ii. Types of Cocoon

Usually a normal tasar cocoon is closed, single shelled and unipedunculate structure with a single pupa inside. The shell is nonflossy and mostly reelable. The anterior dark brown peduncle has an elongated, stiff stalk and distal loop for suspending firmly round a twig. The cocoons are provided with darker veins reticulating across their outer circle. Considering all the parameters such as pupa, shell, peduncle, loop and stalk, cocoons may be classified as follows (Fig. 3.16).

1. Cocoon on Pupal Parameter

Although cocoon with a single pupa is very common in nature but two or three pupae inside the cocoon are seen which are termed as *bipupate* or *Doubled cocoon* and *Tripupate* or *tripled cocoon* respectively. Some cocoons with prepupal stage are immatured cocoons. All the deformed cocoons belong to this category (Fig. 3.26). Bipupate and tripupate may have one or more peduncles.

2. Cocoon on Shell Parameter

In nature, cocoon with a single shell is quite common which is known as *unishellate* or *monoshellate* cocoon. But rarely the combination of two, three or more cocoon shells are noticed which are termed as *bishellate, trishellate* and *multishellate* cocoons respectively. All the unipupate, bipupate and tripupate cocoons are always with a single shell and these are unishellate cocoons. Bishellate, trishellate and multishellate cocoons are jointed and called as double, triple and multiple cocoons respectively.

Broadly they may be categorised as *jointed cocoons*. These cocoons show various complexities in their stalk and loop associations which may be jointed or independent in various ways. Moreover, one or more pupa are also seen in these shells showing complex and anomalous conditions.

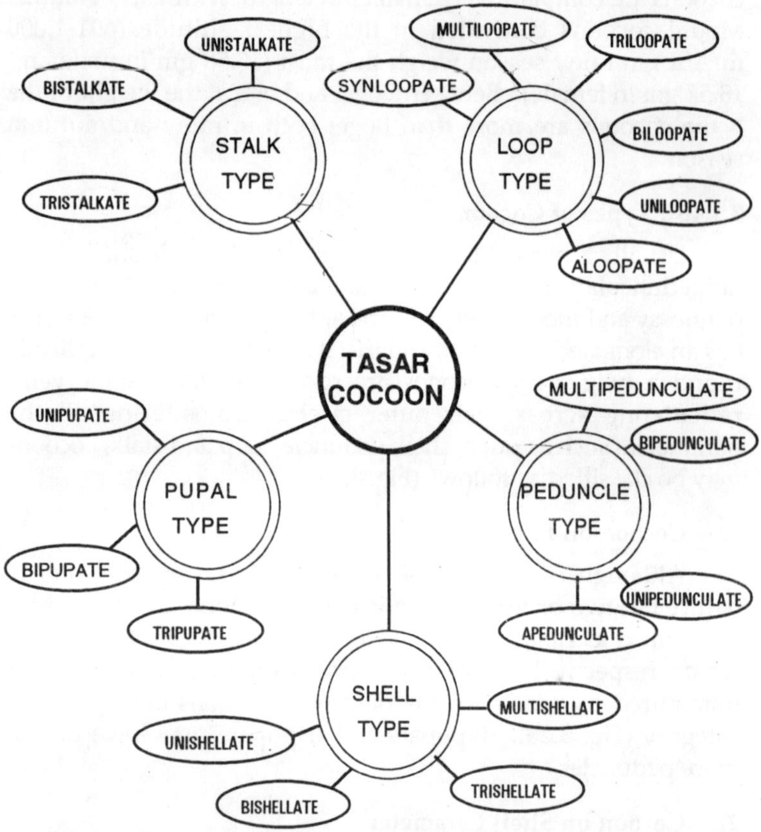

Fig. 3.16 : Typology of Tasar Cocoon

3. Cocoon on Peduncle Parameter

Cocoons with single peduncle called *monopedunculate cocoon* are very prevalent in wild state. But some cocoons without peduncles have been noticed which may be called as *apendunculate cocoons*. Cocoons with two peduncles called *bipedunculate cocoons* are also formed. Bishellate, trishellate and multishellate cocoons exhibit either converging or diverging peduncles.

4. Cocoon on Loop Parameter

Each cocoon with a loop is an integral part of the cocoon since this is a vital structure to hold the cocoon firmly around a twig of the food plant. These cocoons with a single loop are known as *monoloopate* or *uniloopate* cocoons. Some cocoons without a loop have been observed which are known as *aloopate cocoon*. Similarly *cocoons* with two, three or more loops are also noticed which are coined as biloopate, triloopate and multiloopate cocoons respectively (Plate 13). In bipupate, tripupate, bishellate, trishellate and multishellate cocoons fusion and combination of loops are seen which are termed as *synloopate cocoons*.

5. Cocoon on Stalk Parameter

Two types of cocoons on the basis of the length of the stalk have been noticed such as long stalk and short stalk respectively. One, two or three stalk condition in unipupate and bipupate cocoons are observed which are termed as *unistalkate, bistalkate* and *tristalkate* or *multistalkate* cocoons. In bishellate, multishellate, bipupate and tripupate type of cocoons, the stalks are sometimes combined or fused being termed as *synstalkate cocoons*. This *synstalkate* type is further of two types such as *holojointa* and *parajointa* stalked cocoons. Stalks when fused completely in entire length it is called as *holojointa* type and when fused partly in its length it is called as *parajointa* type. The parajointa stalked cocoons may be further classified into three types such as *epijointa, mesojointa* and *hypojointa* depending on the joining portion of the stalks of the cocoons.

6. Fused Cocoon

One pupa per cocoon is the regular phenomenon in silk insects. But the occurrence of two pupae in a single cocoon is not uncommon in tasar silk insects as in mulberry and eri where doubling of pupae occurs through joint spinning of a single cocoon by two age old larvae in association that pupate inside it. This type of cocoon is known as *fused cocoon* or *doubled cocoon* or *amalgamated cocoon*. But this is not a double cocoon which is a misnomer. The double cocoon is formed due to the strong association of close contact of two individual cocoons, having one individual pupa in each cocoon. But in contrast, in the doubled cocoon the cocoon shell spun by two larvae is completely fused to form a single cocoon. So the double cocoon may be called as *jointed cocoon, combined cocoon* or *associated cocoon* in which two cocoons are in close association rather than fusion. Individual cocoon with two pupae is more appropriate to be called doubled cocoon instead of double cocoon in which doubling of pupae and cocoon shell is done by two final stage larvae in association with complete fusion of the cocoon shell.

3.1.4.iii. Sex Pattern in Doubled Cocoons

A cocoon with single pupa is of usual occurrence in all sericigenous fauna. But out of different types of association among pupae, cocoon shells and peduncles, close contact of the outer shell wall of two individual cocoons is characteristic where doubling of pupae occurs through joint spinning of a single cocoon by two age old larvae in association that pupate inside it. The sex association (Sex ratio) in doubled cocoons reveals that male-female (M-F) association is more frequent than that of the same sex. The female-female (F-F) association is more frequent than male-male (M-M) association. The doubled cocoon formation is accidental which is probably induced by some physiological, physical or ecoclimatic stress during the last phase of larval life. However, this is an anomalous behaviour of the last instar due to some unknown factors or may be due to random encounter of the larvae in which one or both the partners are under some stress for which they require to live in close association rather than to live in isolation. But the higher frequency of male-female (M-F) association in doubled cocoons is obvious on the basis of universal attraction between opposite sexes.

3.1.5. IMAGO

The imago is a large beautiful moth with cylindrical body in the middle and tapering end posteriorly. The female moth has greater dimension, diameter of the fenestra and expanded wing than the males. Among the various colour groups in male sex, brown coloured moths have slightly wider wing expanse than the brick-red moths. Similarly, the female moths of sepia colour have the widest fore-wing expanse, body dimension and fenestra among all the three colour groups of moth. However, there is not much difference in wing expanse between sepia and yellow. Pink coloured moths have the shortest wing expanse. Similar is the trend in body dimension and diameter of the fenestra. The body colour of the female moths varies from cadmium, yellow through pinkish or brown. A particular specimen is never a mixture of different colour but of one tone over all. In case of male moth the body colour ranges from brick-red to brown, khaki and even blackish and of shades of such colours. The head is more or less sunken and the antenna is bipectinate in both the sexes but the pectination in males is stronger than females. Maxillary palpi are absent in both the sexes. Wings are broad and strong. The fore-wings or primaries are larger than the hind-wings, the latter are almost rounded. Almost central to the wing occur the more or less circular large transparent fenestra which are prominent black rings more or less of circular shape. Each fenestra is crossed by a vein approximately at right angle to the costa. Body and wings are thickly clothed with hairs and overlapping scales. Frenulum is atrophied and there is basal widening of the humeral area of the hind-wing.

Head

The head is relatively small in comparison with the body, hypognathus hemispherical and mostly covered by the eyes (Fig. 3.17A). It is densely covered with hairs and both in males and females 0.5 cm in length. The head is about 0.2 cm broad across the compound eyes. The greater part of the head is formed by the epicranium which carries a pair of large and globular compound eyes on its lateral part. Eyes are glabrous and ocelli are absent. The head carries a pair of bipectinate antenna posterior to the compound eyes. The anterior region of the head is occupied by a more or less quadrangular sclerite, the *frontoclypeus* (FC), there being no clear demarcation from the epicranium. This sclerite measures 0.3 cm in length. In between the frontoclypeus and the compound eyes, there is a pair of narrow genae on either side. The mouth parts are rudimentary (Fig. 3.17A-E) and located in the anterior most region of the head. Each antenna is set in a membranous area of the cranial wall known as the *antennal socket*, the rim of which is reinforced by a sub-marginal ridge. The external groove of this ridge is the *antennal sulcus*. The posterior concave surface (Fig. 3.17B) of the

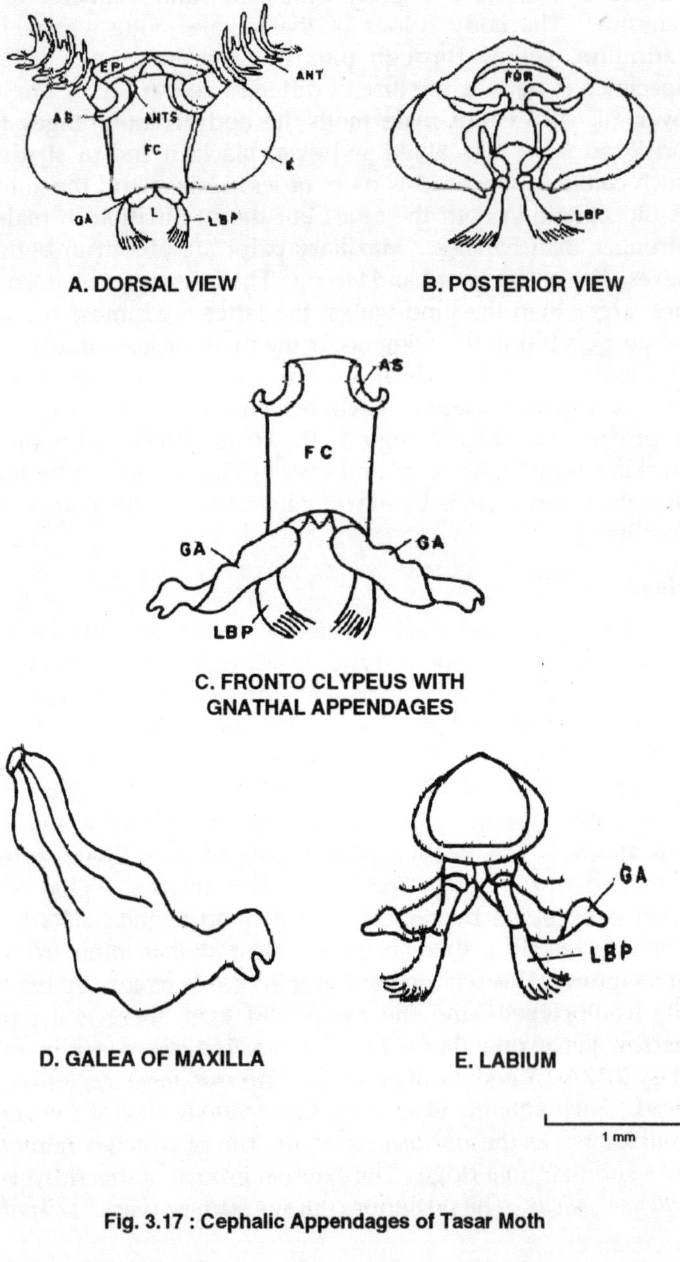

Fig. 3.17 : Cephalic Appendages of Tasar Moth

head capsule carries the occipital foramen. The latter is closely surrounded dorsolaterally by the postoccipital sulcus which cuts of a narrow postocciput arching over the foramen. The postocciput is slightly widened out at its lower extremities on the sides to form the occipital condyles for articulation with the cervical sclerites.

Mouth Parts

The gnathal appendages (Fig. 3.17C) are rudimentary which occupy the anterior most region of the head capsules. These comprise of (i) an unpaired labrum (ii) a pair of mandible (iii) a pair of maxilla and (iv) a pair of second maxilla or the labium. All these structures, however, are greatly reduced.

(i) **Labrum :** The labrum (L) is a narrow membranous sheet hinged to the frontoclypeus (Fig. 3.17A). The lateral hinge is not distinct. The amount of reduction, the labrum has undergone, clearly indicates that it can hardly be of any effective use and is, therefore, merely of vestigial nature.

(ii) **Mandibles :** Each mandible is a typical structure. Its base as usual has a slightly condylar thickening at its inner edge and a long tendon is attached to its outer angle. Its feeble sclerotisation shows its vestigial nature.

(iii) **Maxillae :** Maxillae are greatly reduced and the maxillary palpi are absent. The stipes is articulated laterally with the head capsule. The proximal end of the stipes has a small chitinous structure which represents the cardo. The proboscis is abortive and non-functional. The two galea (GA) remain separate (Fig. 3.17D), clearly indicating that no food can be imbibed. The lacinia is absent. Both the cardo and stipes are together lodged into a shallow concavity in the head capsule.

(iv) **Labium :** The labium is reduced to a small plate on the ventral part of the mouth. It is differentiated into a membranous postlabium and sclerotised prelabium. There is a pair of three segmented labial palp (LBP) (Fig. 3.17E), the third segment being the longest and is densely covered with hairs and it appears like a brush.

Antennae

The antenna is bipectinate in both the sexes. In male the antenna (Fig. 3.18A) is more strongly bipectinate than in the female and measures 1.5 cm to 1.6 cm in length comprising of 40 segments. The flagellum (Fig. 3.18B) comprises of 37 segments. The plumules are nearly in a flat plane making the antenna a plane of about 0.7 cm across narrowing to each end and wider at the centre of flagellum. Each joint bears four plumules which arise almost from the dorsum. The regions of pectination of the joints are dorsal rather than lateral and appears to be regularly placed, nevertheless the alternate branches differ, the distal branch is bent more dorsally whereas the basal is bent more ventrally. The basal ones are a little darker in colour than the rest and terminate more bluntly. Both are cylindrical. The basal is little thickened at the end and terminating in a point. Each branch bears sensory hairs of about 0.2 mm long. The shaft carries a few similar but shorter hairs. Both the shaft and the branches are slightly yellow—ochre in colour and free from scales.

The female antenna (Fig. 3.19A) is about of the same length as that of the males and has the same number of joints while the sensory hairs are absent unlike in the males. It is also yellow-ochre in colour. The pectination of female antenna is represented by processes of which the basal ones are longer (Fig. 3.19B). The distal pectination is merely a thickening and projection of the lateral margin of the end of the joint but are well developed at the middle of the antenna. The upper or outer processes, however, of both series are much smaller than the lower. The upper basal and lower basal have two soft terminal bristles.

Pterothorax

The pterothorax (Fig. 3.20, 21, 22) is composed of the wing bearing segments mesothorax and metathorax which differ structurally from the prothorax. The modification chiefly affects the terga, in a lesser degree the pleura and least the sterna.

Mesothorax

It is the largest of the three thoracic segments and is highly complicated in structure. The mesonotum covers the major part of the dorsum of the thorax and is composed of the alinotum

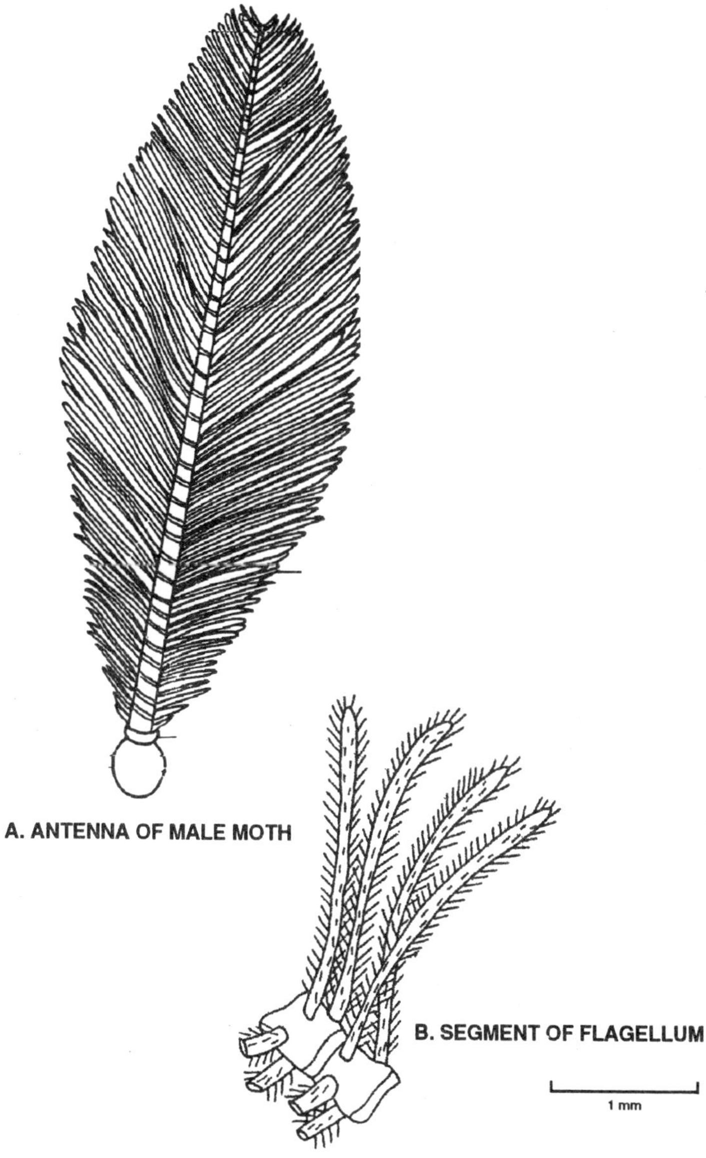

Fig. 3.18 : Antenna and Flagellum Segment of Male Moth

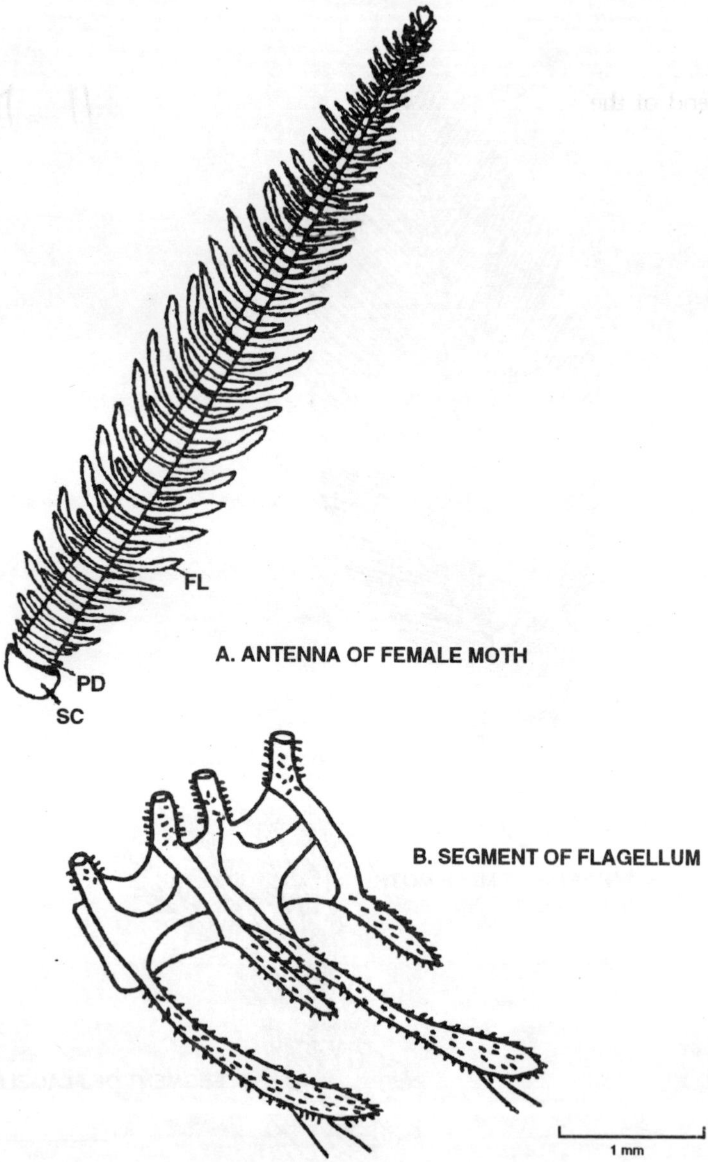

Fig. 3.19 : Antenna and Flagellum Segment of Female Moth

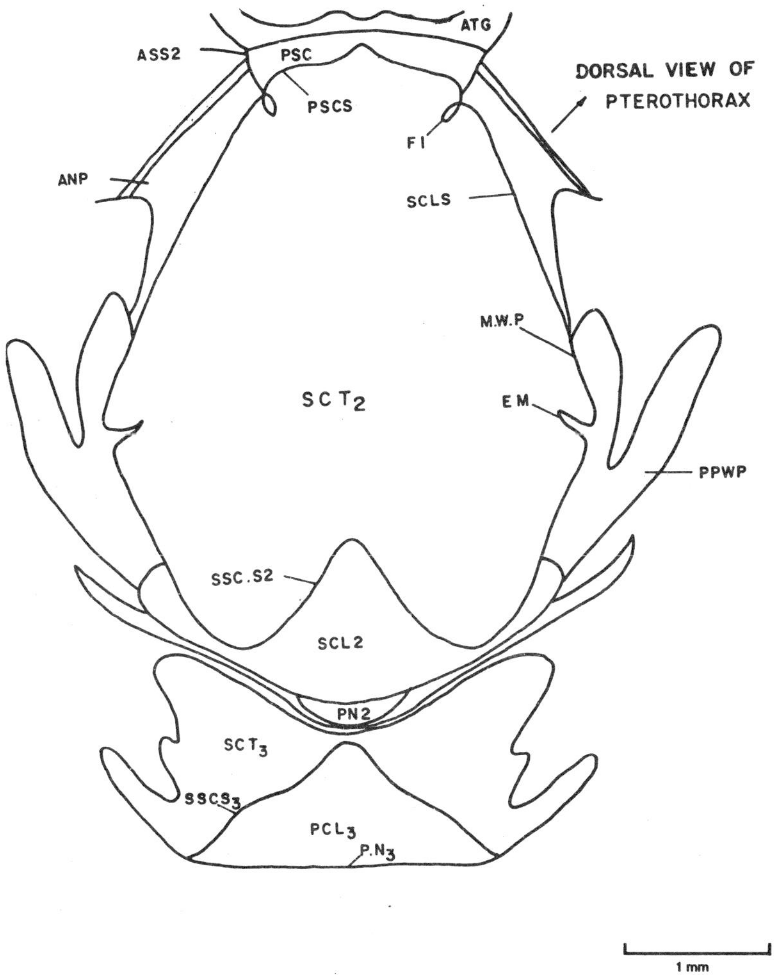

Fig. 3.20 : Dorsal View of Pterothorax of Tasar Moth

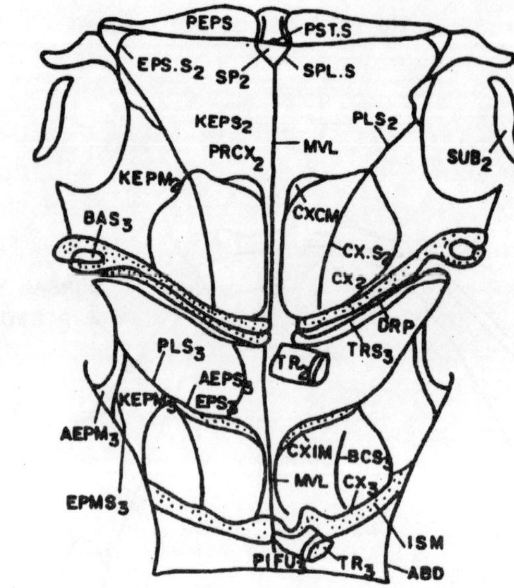

Fig. 3.21 : Ventral View of Pterothorax

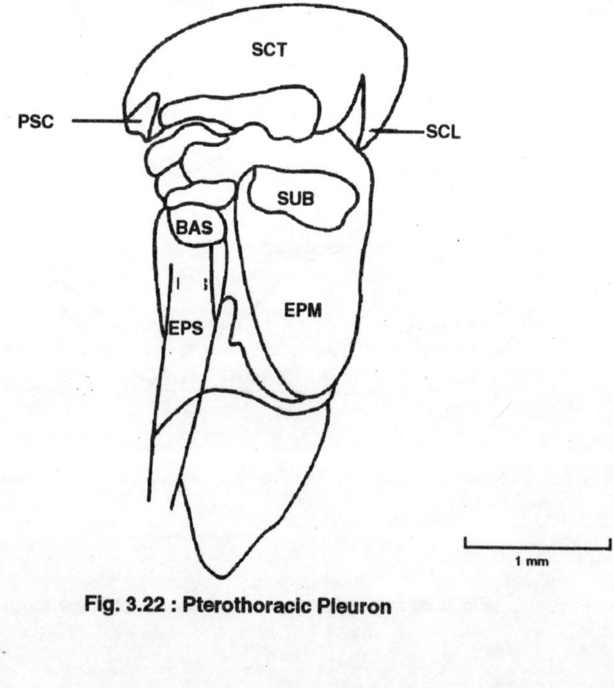

Fig. 3.22 : Pterothoracic Pleuron

and the postnotum. The former is distinguishable into four areas : (i) acrotergite (ATG), (ii) prescutum (PSC), (iii) scutum (SCT_2) and (iv) scutellum (SCL_2). The acrotergite is a narrow marginal plate with a median notch. The posterior end of the prothorax is linked into this notch. The posterior limit of this region is marked by the antecostal sulcus (ACS_2). The prescutum is also a narrow sclerite which lies sunken between the acrotergite in front and the scutum behind and eventually invisible externally in the dorsal view. Anteriorly it is bound by the antecostal and posteriorly by the prescutal sulci. Each lateral part of the prescutum corresponding to the prealar has sunk into the form of a peglike *prealar process.*

The scutum lying behind the prescutum forms the major portion of the mesonotum. It is strongly arched and is narrow anteriorly and relatively broad posteriorly. From each lateral margin of the anterior half of the scutum there is a projection some what triangular in shape, the anterior notal wing process (ANP). The base of the process is marked by a sulcus, the scuto- lateral sulcus (SLCS) while its margin is marked by another sulcus the marginal sulcus. Both the sulci form the strong internal ridges. The anterior and posterior ends of the scuto - lateral sulcus are marked by a fissure (FI) in the scutum. The posterior half of the scutal margin gives rise to another process, the *posterior notal wing process* (PNP). Between the two notal processes there is the scutal emargination (EM) in the form of fissure. The presence of the distinct scutal emargination clearly indicates that the median wing process definitely belongs to the posterior notal wing process.

The scutellum is demarcated from the scutum by an inverted V shaped *scuto - scutellar sulcus* (SSC_2S). Each end of the sulcus is located at the base of the posterior notal wing process. Internally the sulcus forms a strong ridge which imparts strength and firmness to the arched portion of the mesonotum. The posterior portion of the scutellum is bent downwards to be connected with the underlying mesopostnotum (PN_2) through inter segmental membrane while the basal corners are continuous with the axillary cords of the forewing. The mesopostnotum is considerably enlarged and extended forwards beneath the mesoscutellum carrying antecostal sulcus and midphragma with it belonging to the metanotum.

Mesopleuron

The mesopleuron does not differ fundamentally from the prothoracic pleuron but secondary differences between the two are considerable on account of the degenerative tendency of the prothoracic pleuron. The mesothoracic pleuron is typically a continuous sclerotic area in the lateral wall of the mesothoracic segment surrounding the base of the coxa dorsally, anteriorly and posteriorly. It is divisible by a prominent pleural sulcus and its ridge into a ventral episternum and a dorsal epimeron. The strong pleural ridge is fused dorsally with the transverse ridge and ventrally with the coxal ridge of the mesocoxa. The episternum is larger than the epimeron. It projects anteriorly into a sulcus, the preepisternal sulcus that cuts off an anterior lobe, the preepisternum. The inner ends of preepisternal sulci of the two sides are continuous with the outer ends of a small *presternal sulcus* (PSTS) so as to form a long transverse sulcus having a prominent internal ridge.

The episternum itself is divisible by a small episternal sulcus into two regions, a ventral *katepisternum* ($KEPS_2$) and a dorsal *anepisternum* (Fig. 3.4). The katepisternum is a large plate and its ventral portion, anterior to the subcoxal cavity is the procoxale. The postero-lateral margin of katepisternum, limiting the subcoxal cavity anteriorly is strengthened by a ridge extending from the ventral end of the pleural ridge. The anterior ends of the two sternopleural sulci are still free to bound the two sides of a small triangular area, the anterior margin of which is limited by a small presternal sulcus. The mesofurcal pit is situated at the posterior extremity of the midventral line and is the external remnant of the *mesofurcasternum*. The pit is composed of a stalk with two furcal arms. Each furcal arm is flat and roughly triangular structure. Posterolaterally, each arm is fused with the postero-dorsal wall of the pleuron.

Metathorax

The metanotum is about one third of the mesonotum and is differentiated into (i) scutum (SCT_3), (ii) scutellum (SCL_3) and (iii) post-notum (PN_3). The prescutum is absent.

(i) **Scutum :** The scutum constitutes the major portion of the notum. Due to the anterior extension of the apex of the scuto - scutellar sulcus, the scutum is divided into right and left parts (Fig. 3.20). These parts are triangular with their bases making the lateral margin of the notum and their apices directed towards each other. Laterally, each part has a lateral sulcus which runs parallel to the lateral scutal margin. Each lateral margin of the scutum is hinged against the mesial margin of the first axillary sclerite. The anterior notal wing process which is migrated to the posterior, is considerably reduced and articulated with the posterior end of the first axillary sclerite. The posterior notal wing process is located immediately behind the anterior notal wing process and articulates with the proximal end of the third axillary sclerite. The *scuto-scutellar sulcus* ($SSCS_3$) has the shape of an inverted V with its ends situated at the bases of the posterior notal wing processes and the apex extending up to the anterior margin of the scutum.

(ii) **Scutellum :** The scutellum is a triangular sclerite, anteriorly demarcated by the scuto-scutellar sulcus. The apical portion of this sclerite extends up to the anterior margin of the scutum, while its posterior portion is bent vertically downwards and forwards to be connected to the *metapostnotum*.

(iii) **Postnotum :** The metapostnotum is overlapped by the posterior position of the metascutellum. Its posterior limit is marked by the antecostal sulcus which is infolded to form the postphragma.

Metapleuron

The metapleuron is drawn into the sternal region due to the invagination of the metasternum and the two pleura meet along the mid-ventral line. Each metapleuron is divisible by the pleural sulcus and its ridge into *episternum* and *epimeron*. The episternum is anteriorly demarcated by the preepisternal sulcus and thus, a narrow marginal preepisternum is cut off. The preepisternal sulci of the two sides are fused to form the transverse sulcus. The two preepisterna are fused to form a narrow prepectus. The episternum is divided by the episternal sulcus into katepisternum (KEPM) and anepisternum ($AEPM_3$). The procoxale is present anterior to the subcoxal cavity. The latero sternite contains the ventral and coxal articulation. The epimeron is also divided into *katepimeron* and *anepimeron*.

The portion of the katepimeron that connects the pleuron with the sternum is the postcoxale. The antero-dorsal portion of the anepisternum and the anepimeron together forms the small pleural wing process. The basalare is borne anteriorly by the episternum while the subalare is carried by the epimeron.

Metasternum

Like the mesosternum, the metasternum is infolded to form a keel like structure. Due to such infolding the pleura of the two sides are drawn close together to meet along the mid-ventral line. The metafurcal pit is located at the posterior end of this line. The pit leads into the metafurca.

Legs

There are three pairs of legs borne by the three thoracic segments. The legs are long, slender and densely hairy (Fig.3.23). Each leg consists of a relatively short *basipodite* and a longer telopodite. The basipodite has two segments, coxa (CX) and *trochanter* (TR) and the telopodite has four segments that is femur (FE), tibia (TI), tarsus (TR) and pretarsus (Fig. 3.23A, B, C).

Basipodite

The coxa is the basal joint of the leg and has the form of a truncated cone. Its anterior surface is greatly elongated as compared with the posterior surface, thereby making the rim of the coxa large and oblong. About three fourth of the boundary of the subcoxal cavity is occupied by the margins of the anterior wall of the coxa and only the remaining one fourth by the posterior wall. The coxal corium is obliterated in the anterior region and reduced to a very narrow strip in the dorsal and ventral regions. The basicoxite is almost obliterated in the anterior region. Marginally it has three *apodemes*. The basicostal sulcus (BCS) (Fig. 3.23F) forms a low basicosta (BC). Ventrally the rim of the coxa (CXR) is discontinuous and in the gap is placed the *sternocoxale* (STCX). The trochanter forms the second joint of the leg. It consists of very short, more or less triangular sclerite freely movable by a horizontal hinge of the coxa but more or less fixed to the base of femur. The articulation between the coxa and trochanter serves the principal movement of the leg base.

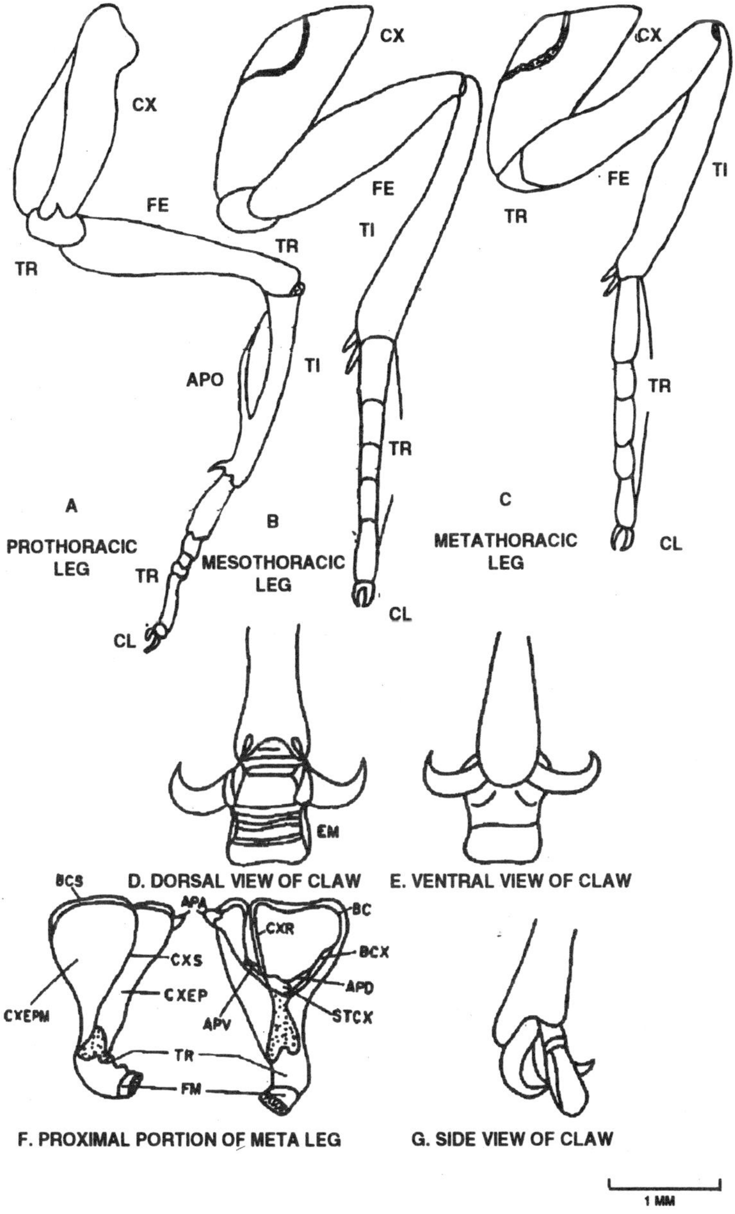

Fig. 3.23 : Leg and its Appendages of Tasar Caterpillar

Telopodite

The segments of the telopodite are relatively longer. The femur consists of a single segment. The tibia has only one segment and is usually a little more slender than the femur. The anterior tibia is comparatively short and is provided on its inner surface with epiphyses (APO). Hind tibia in the males is provided with one spur whereas in the female it is provided with two comparatively smaller spurs. The tarsus is subdivided into five tarsomeres. The first tarsomere is the largest and distinguished as metatarsus.

Pretarsus

The terminal part of the telopodite is the pretarsus. It consists of a paired claw (CL) (Fig. 3.23D,E,F).

Wing

There are two pairs of wings, the anterior pair being larger in size than the posterior. The wing span and forewing is larger in female (Fig. 3.25A). Both the pairs of wings are densely covered by different types of scales (Fig. 3.26a to f). The scales adjacent to the veins are short whereas those away from the veins are long and slender. The prominent postmedian line is red with a white line on its border in both the sexes. In male the antemedian line is usually black or dark drown and is bordered on the inner side with a white line. But in female, the colour of the antemedian is variable that is grey, deep yellow, brown or black. Along with the postmedian line runs a dark brown oblique line with a white inner border. Almost centre to the wing, the *fenestra* or *ocellus* (Fig. 3. 24, 25) is present having black coloured rings with a hyaline area in the middle and surrounded by different coloured zones. It is of various shape. The area between the post-median line and fenestra is darker.

Wing Coupling Apparatus

The adult *Antheraea paphia* exhibits amplexiform type of wing coupling mechanism. The frenulum is atrophied. The humeral lobe of the hindwing is enlarged and strengthened by the development of a short humeral vein and the hindwing is projected far beneath the forewing. As a result, the fore and hindwing are maintained together owing to their close

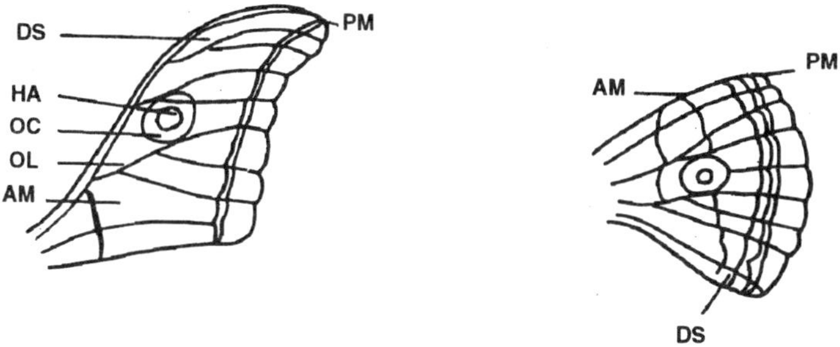

A. FOREWING B. HINDWING

Fig. 3.24 : Wings of Male Moth

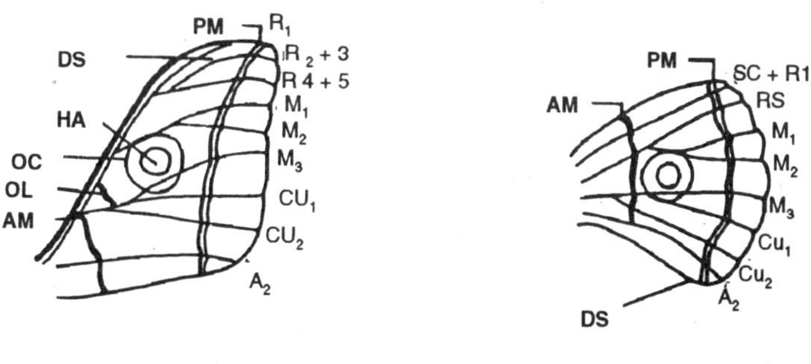

A. FOREWING B. HINDWING

1 Cm

Fig. 3.25 : Wings of Female Moth

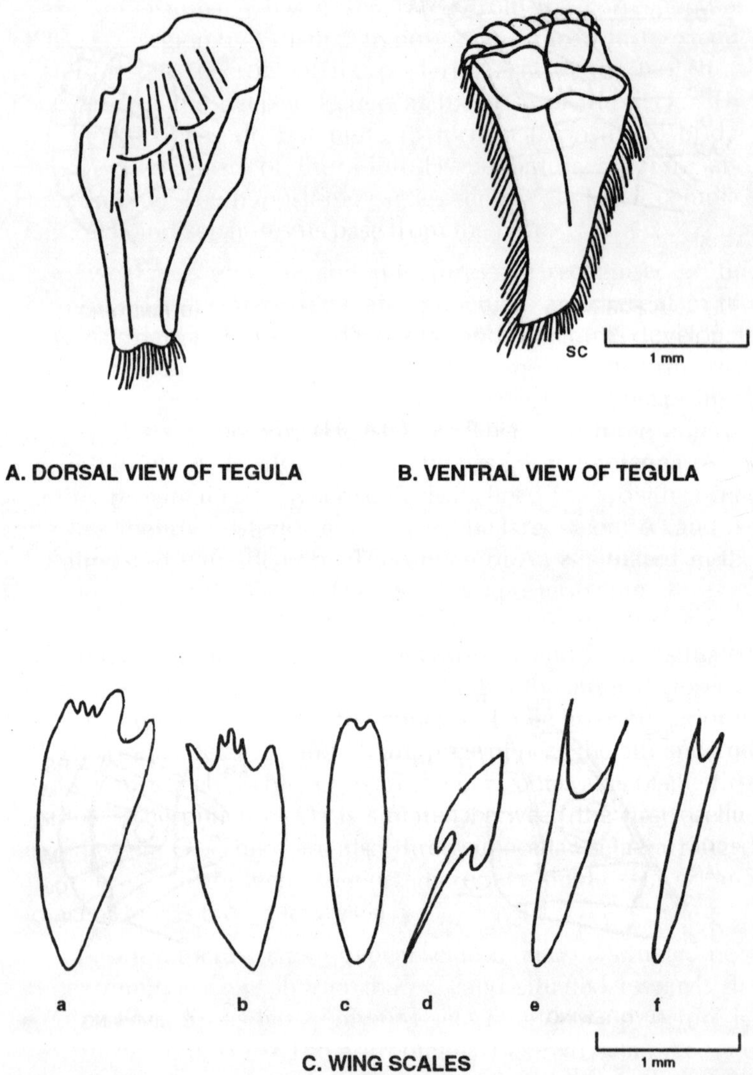

Fig. 3.26 : Tegula and Wing Scales of Tasar Moth

overlapping. The humeral angle of the hindwing is largely developed and usually strengthened by two deep furrows, the bottom of which is thickened so as to appear as a humeral vein.

Wing Venation

Forewing : The subcosta (SC) is well developed, separated, unbranched and does not reach the apical margin. Radius is provided with three branches, all arising from a common stalk (Fig. 3.24A, 25A). These branches are R_1, R_{2+3} and R_{4+5}. R_1 arises from the stalk above and outside the discoidal cell. The discoidal cell is closed. The stalk R_1, R_{2+3} and R_{4+5} arise from before the angle of the discoidal cell. The stalk is smaller than the branches. The median M_1 and M_2 are forked and arise from the apex of the discoidal cell. M_3 is present. The cubitus Cu_2 arises from the lower angle of the cell. Out of anal veins, A_2 is present but A_4 is absent. Vein M_1 is joined to the radius by a cross vein r - m. M_2 arises mid way between cubitus and radius leaving the latter apparently three branched. The fenestra is divided by the discal vein.

Hindwing : In hindwing, the postmedian line is restricted. The antemedian line is wavy and devoid of a white border. The oblique line is absent. Sc + R_1 and Rs are united at the point of origin, but separate for the most of the wing length. M_1 arises from the angle of the discoidal cell. Base of M_2 is closer to M_1 than to M_3. Base of M_3 is closer to that of Cu_1 than to that of M_2. Cu_2 arises from the lower angle of the discoidal cell. A_2 is present. M_1 is joined to radius by the cross vein r - m. The fenestra or ocellus is divided by a discal vein. The discoidal cell is closed (Fig. 3. 24B, 25B).

Abdomen

The abdomen is elongated and conical. There are ten abdominal segments which are densely hairy. The visceral segments of the abdomen are simple in structure and differ but little from one another. They do not bear any appendicular organs. The first abdominal segment is reduced (Fig. 3.27C) and its sternum is wanting. The rest of the visceral segments are simple and possess similar terga and sterna.

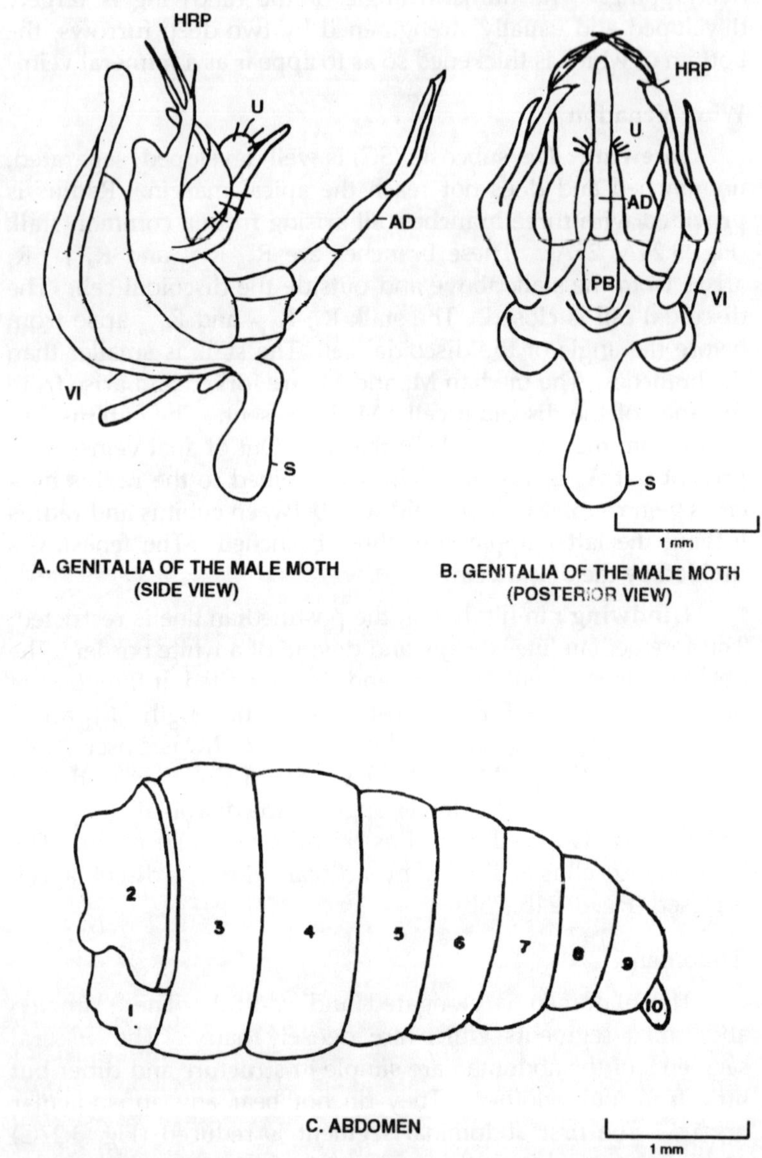

Fig. 3.27 : Abdominal Appendages of the Moth

External Genitalia

Male moth : Both in males and females, the genital complex is formed by 8th, 9th and 10th segments. The 8th segment is the least modified of the three and forms a protractile base for the rest of the genital apparatus. It has a well developed tergum the *menis dorsalis.* The sternum is also sclerotised and forms the *menis ventralis* (Fig. 3.27A, B). The segment 9 is greatly modified with distinct tergal and coxa-sternal areas of sclerotisation. The tergum is known as the *tegumen* and encircles the apex of the body. The coxa-sternal region is known as the *vinculum,* the sternal region of which is extended forward in a deep inflection below the preceding segment known as the *saccus* (S).

The periphalic organs of the genital complex consist of a pair of claspers of valves hinged to the vinculum, forming a very prominent part of the male genitalia. These are more or less flattened triangular lobes covered by hairs. The valves are divided into three parts : *costa, valvula* and *sacculus* which are all fused together. Besides, the valves are clothed with long marginal spines. Borne on the coxopodite area of the vinculum there are a pair of spinelike prehensile organs known as the *harpes* (HRP). They are present in the inner portion of the claspers. The tergum of the 10th abdominal segment is modified to form the *uncus* which is fused at the caudal end of the tegumen. It is more or less spherical in shape and provided with numerous bristles. The anal tube passes below the uncus and there is no trace of the gnathos. The phallic organs consist of an intromittent organ, the *aedeagus* which is enclosed in an eversible sheath. At the point of attachment of the sheath with the body, there is a sclerotised support known as *juxta.* The aedeagus is not armed with spines (Fig. 3.27).

Female moth : The female genitalia is eversible and retracted into the 7th segment. The genital complex is formed by the 8th, 9th and 10th segments. The 7th segment has hardly undergone any modification. The 8th segment is highly modified (Fig. 3. 28 A, B). The tergum is large, whereas the sternum is greatly reduced. The posterior sternal margin forms a broad **V** shaped structure. The tergum and sternum of the 8th segment are fused together to form more or less a ring shaped structure. The tergum at its point of union with the sternite is produced

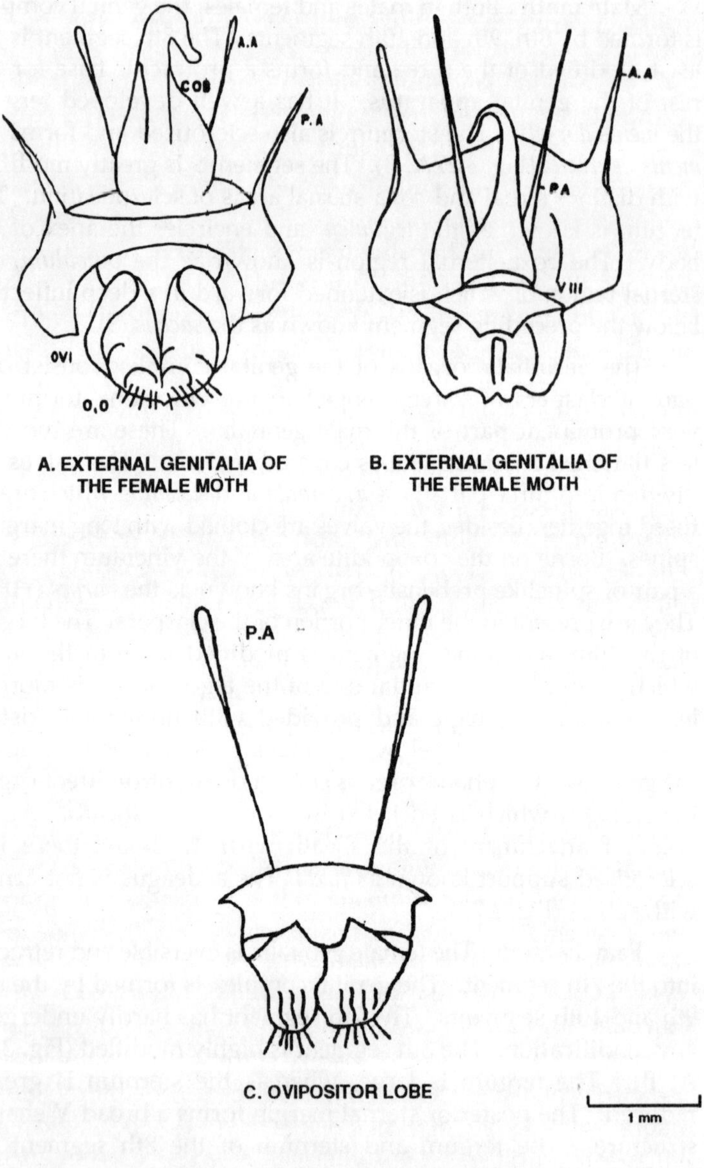

A. EXTERNAL GENITALIA OF THE FEMALE MOTH

B. EXTERNAL GENITALIA OF THE FEMALE MOTH

C. OVIPOSITOR LOBE

Fig. 3.28 : Genitalia of Female Tasar Moth

into a pair of highly chitinized rod-shaped structures, the *anterior apophyses*. The bursacopulatrix (COB) is a tube like structure which opens by means of an *ostium bursa* at the midventral margin of the 8th sternum. The sternum at this region is very small and forms the genital plate. The tergum of the 9th segment is very much modified while the sternum remains largely unsclerotised. The tergum is fused posteriorly with the 10th segment which forms a pair of ovipositor lobes and anteriorly it is produced into a pair of chitinized rods the posterior apophyses (PA) which extends up to the end of the 8th segment. The 10th segment is greatly modified to form a pair of more or less triangular sclerotised lobes, the *ovipositor* (OVI) lobes (Fig. 3.28C) which are clothed with numerous setae and form the posterior most element of the genital complex.

3.2. Rearing

Various ecoraces of tasar silk insect such as Godamodal, Nalia, Bogei, Sukinda and Daba exist. Godamodal and Nalia are the wild tasar ecoraces known as *Antheraea paphia* L. whose first generation of semidomesticated variety is the Bogei. Sukinda and Daba ecoraces are domesticated in nature which belong to *Antheraea mylitta* Drury. Rearing of domesticated variety is easier in comparison to wild variety. Outdoor rearing exposes tropical tasar silk insect to unfavourable weather conditions and attack by various pests and predators. These account for losses of more than 50% mainly during the early instars. Besides, improper handling of larvae and the faulty selection of rearing site and plantation, for example, cause serious harm to the health and vigour of the population, which is reflected in a higher rate of mortality from disease. Since losses can be substantially minimised by a proper rational approach, rearing practice needs systematic attempts. The fate of the crop largely depends on choice of rearing site and food plants, brushing, supervision and maintenance of larval population.

3.2.1. Rearing Site and Food Plants

The traditional tasar rearers usually utilise the sparsely distributed food plants on the bundhs of paddy fields or any patch of land with some host plants as a conventional practice. Waterlogging in the fields during rains raises the relative humidity. Also silk insect excreta and dead larvae which fall on water putrefy and pollute the environment. During the dry winter months, the scattered plants fail to maintain the atmospheric humidity at the desired level. All these unfavourable conditions promote the outbreak of diseases. So it is desirable to avoid low-lying areas, which are susceptible to waterlogging. Secondly, thick patches of food plants of 3.0 - 3.5 m tall are ideal. Thirdly, same bushes must not be utilized for two successive crops in a year.

3.2.2. Maintenance of Rearing Site

Before the onset of rearing, the site and the surrounding area require to be cleaned of weeds, which might otherwise induce pest and predator attacks. Apart from removing the unrequired foliage, care should be taken to free the bushes of insects particularly nests of white ants and ants. A band of straw with a little ash should be tied around the trunk to prohibit the downward movement of the larvae. The trunk base should be encircled with a thin band of bleaching powder to prevent attack by white ants, ants and other insects.

3.2.3. Foliage Quality

The quality of the leaves in relation to the larval age is a major determinant of health and vigour. The younger larvae thrive on juicy and tender leaves, but pale green foliage is harmful. The subsequent instars require medium to mature leaves which are deep green with tough cellulose content. Outdoor rearing cannot ensure provision of the proper quality of leaves during the different instars. Nevertheless, this can be achieved to a degree through various cultural operations and selective utilisation of food plants. For different instars suitable quality of leaves can be given only during indoor handling and rearing operation.

3.2.4. Brushing

After the larvae hatch out, the placing of the larvae on the leaves is a key step of rearing. Cultivators tie the leaf cups containing the eggs on the bushes for hatching as a traditional practice. In this case, both the developing embryo and the newly hatched larvae are thereby exposed to fluctuating temperature, humidity, rainfall, storms and various other environmental changes which result in poor hatchability and significant losses of larvae.

3.? ᄃ. Supervision and Maintenance of Larval Population

Traditional rearers brush the larval population too densely without keeping the quality of foliage in view. This high density decreases effective yield owing to the higher mortality due to disease and also adversely affects the economic character of the cocoon. Frequent direct handling of the larvae contaminates the population causing considerable injury to their health. Moreover, traditional rearers although are advised to handle the larvae by formalinised hand but they avoid using formalin due to their negligence and unscientific approach. It is always desirable to transfer the larvae once or twice by cutting of the small branches bearing larvae and attaching them to unused food plants. Another advantage of this system is light pruning of the plants.

Moulting and spinning larvae require special care and attention; the former should never be disturbed and the latter require enough foliage to form the hammock, a prerequisite structure for cocooning. Lack of sufficient foliage in plants affects the larvae to form cocoon suitably. Dead larvae hanging from the foliage of the bush or fallen to the ground needs to be collected every morning and late afternoon. A sample of dead larvae should be examined for *microsporidiosis.* All the dead larvae should be buried or burnt at a great distance at least away from the rearing site. Secondly, larvae as well as foliage having symptoms of microsporidiosis should be destroyed. Those showing other diseases should be reared in a separate site. Rearing appliances and the hands of the rearers or field staff should be disinfected with dettol water after every contact. Cocoons should be harvested after six or seven days. The branches are cut and the cocoons are pulled off the twig by breaking it near the ring.

3.2.6. Modified and Improved Rearing Technique

Outdoor rearing technique causes a series loss of population particularly during the first instar. But attempt to total indoor rearing practice have not been successful so far because of several difficulties like indoor conditioning and the availability of natural facilities. So to overcome these hurdles, controlled rearing techniques have been adopted and practised to protect the early stages. The larvae are reared indoors until the first moult. Brushing is performed according to the improved technique of using only one day's hatchlings and avoiding overcrowding. Care is taken to remove the enclosure (controlled rearing set) as soon as the larvae start settling to moult; otherwise the ecdysis becomes difficult so as to cause heavy mortality. The larvae coming out of moult are allowed to crawl on to fresh twigs and transferred outdoors taking to precautions into notice. The advantage of controlled rearing is that it reduces not only the first instar losses but also mortality by disease during the subsequent stages.

The development of economic plantation of tropical food plants of tasar has promoted the concept of early instar rearing, which is a further improvement on the basic principle of controlled rearing. Rearing up to the third instar is conducted on economic plantation preferably under nylon netting. Although the overlapping twigs allow the larvae to crawl from one bush to the other, the medium size and regularity of the plantation not only permits efficient management, supervision and operation but also minimises losses. Moreover, the larvae are more vigorous as a result of being reared under natural environment with almost no intervention. As soon as the larvae have passed the second or third moult, the twigs bearing them are cut and transferred to forest or block plantation for rearing in the advanced stages. To save time and economise labour in the transfer of early instar-population, the rearing centres should be located as close to the forest as possible. Nearly 4,000 disease free layings (Dfls) can be reared up to the third instar on one hectare of economic plantation.

3.3. Pathology

Diseases are the cause of crop loss every year in each season. Tasar silk insect suffers from four diseases : (i) *microsporidiosis* or pebrine disease (ii) *ployhedrosis* or *virosis*, (iii) *bacteriosis* and (iv) *mycosis* or fungal disease. Among these four diseases, while viral, bacterial and fungal diseases are transmitted orally through ingestion of infected food material, pebrine is transmitted through contaminated food as well as egg cytoplasm. Though tasar insect *Antheraea paphia* is prone to bacterial disease (muscardine), viral disease (grasserie) and pebrine (microsporidian disease), pupal mortality is recorded due to pebrine infection. The pebrine disease forms a permanent scourge in the tropical tasar and hence rigorous steps need to be taken to control the disease since it multiplies and spreads like wild fire. Moreover, due to lack of a convenient method of pebrine detection in grainage it has not been possible to control this particular infection so far. Therefore, a systematic and suitable method has been suggested to detect the pebrine spores.

3.3.1. Etiology, Pathogenicity, Epidemiology and Histopathology

The pathogen causing pebrine disease is a protozoa belonging to the class Sporozoa, order Microsporidia, family Nosematidae and genus *Nosema*. Its life cycle includes three stages such as *spore, planont* and *meront*. The pathogen multiplies extensively in different tissues including digestive tract, fat bodies, malpighian tubules, silk glands, gonads and body wall showing dissimilar lesions. The activity of the host becomes severely impaired due to infection. The presence of the pathogen on the muscles causes distention of the tissues thus making the larvae unable to pass into pupal and adult stage. The proliferation of the meront to produce newly formed spores causes the infected cells of the host to swell, burst and disintegrate leading to the death of the insect. The major factor in the spread of the disease is the rapid multiplication of the pathogen and their dissemination.

The cause and source of infection in tasar silk insect are many. The foremost among them being the transovarial mode of transmission. The source of contamination through corpses of diseased worm excreta, contaminated egg shell, leaves, exuviae, scales and cocoon shell is also quite common. Though the degree of transovarial transmission has been found to vary considerably, contamination within the worms reared on the same plant is caused by the dissemination of the spores in large quantities through the diseased worms on the plants. Ingestion of the spores by healthy worms results in secondary infection of the worms. The tasar insects contracting the disease during fourth or fifth instar are able to form the cocoon and emerge out as adults but the eggs laid by them carry the disease in the egg cytoplasm. The extent of infection due to secondary infection depends on many factors including the percentage of diseased worms in that particular lot, stage of infection and agroclimatic condition.

The spores of *Nosema* invade the mid gut epithelium, longitudinal muscle fibre, epidermis of the integument, ovary, testis, silk gland and fat body. Though the infection is light in parts of the mid gut and muscle fibre, it is quite intense in the fat body. The black spots over the larval body are the outcome of the lesions visible through translucent cuticular layer.

3.3.2. Modern Method of Pebrine Detection

Since pebrine infection is very much common in tasar fauna, an innovative method of detection in early stage is suggested. Pebrine creates the greatest problem in preparation of seed in commercial grainages. The exuviae of primarily infected larvae and larvae infected in the third instar are examined for pebrine spores. The exuviae are crushed in 2 per cent KOH and centrifuged at 4,000 rpm for 10 minutes and sediments are examined under the microscope. The sediments indicate to contain pebrine spores of the pupae were infected with pebrine disease. Another test through microscopic examination of the smear preparation of moths, larvae or pupae confirms the spores as bright oval bodies measuring on an average 5.8μ and 2.3μ in length and width respectively. This procedure effectively reduces the load at the time of actual grainage since this is the easiest and the most suitable technique to detect and screen out the infected pupae before the commencement of grainage.

3.3.3. Control of Pebrine Disease

Due to sporadic incidence of the pebrine disease in tasar zones, a comprehensive and integrated approach is imperative to control the disease. Since it is essential to prevent the disease before it occurs, following measures are suggested for combating the disease :

1. For preparing quality seed, microscopic examination of individual mother moths has to be strictly followed and the eggs obtained from the disease free mother moths alone should be utilised for rearing.

2. Seed cocoons should be procured exclusively from non-infected area and seed area should be earmarked. Before selection of seed zones, the rearing areas should be surveyed and as per the previous rearing performance of at least 5 - 6 crops, the area can be earmarked as seed zone.

3. Cocoon from the infected area should be utilised for reeling and that of noninfected area should be collected for rearing purposes only after seed requirement is fully met with.

4. The flow of cocoons from infected to noninfected area should be banned.

5. Aseptic condition should be maintained during grainage and rearing operation.

6. Grainage houses and rearing huts should thoroughly be disinfected or fumigated after the completion of grainage and rearing.

7. Disease free layings (Dfls) or certified eggs should invariably be disinfected in 5 per cent formaldehyde solution for 5 minutes.

8. The grainage operations should be conducted far away from the rearing field as far as practicable. At least future constructions for the purpose of grainage should be constructed away from the rearing field.

9. The grainage house should be provided with a false ceiling of bamboo mats over which a plaster of mud should be applied to reduce the temperature further. This also helps in proper disinfection of the room. Just below the false ceiling, the grainage house should be equipped with exhaust fans facing each other through the length of the room.

10. At places, where incidence of the disease is high, the food plants should be disinfected thoroughly with a mixture of 2 per cent formaldehyde and 2 per cent bleaching powder solution. The grass under the bushes should be burnt for eradicating the chances of contamination through the soil. These operations should be conducted during the first week of February or on completion of third crop, whichever is earlier.

11. Spraying of 2 per cent formaldehyde solution should be done everyday after completion of the work in the grainage.

12. Moths for egg laying should be kept in a separate room and in no case in the grainages where emergence is continuing, in order to avoid the chances of contamination through scales, dead moths and cut wings.

13. During egg laying, a thin layer of bleaching powder should be sprayed on the floor and then the earthen cups or paper boxes should be kept over it for egg laying.

14. The eggs should be first washed in 0.5 per cent sodium hydroxide solution or soap solution to remove the muconieum layer and the disinfected in 5 per cent formaldehyde solution for 5 minutes followed by a wash in distilled water. The material used for the disinfection of eggs should not be used more than twice.

15. Drying and incubation of the egg should not be conducted in the same building where other grainage operations are being done. This should be conducted in another building where the chances of contamination through scale etc. is ruled out.

16. Earthen cups and other equipments used during grainage should be sterilised before every use. Sterilisation should be conducted in 2 per cent solution of formaldehyde under boiling condition for at least an hour.

17. The area around the grainage house should be kept clean. The practice of raising flower bearing plants and other herbs by the side of the grainage should be avoided as it acts as an abode for the pathogens because invariably after sweeping the grainage house, the refuse containing the contaminated material is likely to be retained over there.

18. During grainage operations, utmost care should be taken to engage a particular person for a specific work, for example, a person engaged in handling the moths should not be allowed to handle the eggs.

19. During examination of moths, arrangement for running water must be made for washing the scissors, mortar and pestle and other required equipments. This should be followed by a wash in 2 per cent formaldehyde solution.

3.3.4. Disease of Food Plant and its Control

Proper growth of the larvae results in proper cocoon quality. But the growth of the larvae depends on the leaf of the food plants. One of the serious seasonal diseases of the Arjun plants *(Terminalia arjuna)* occurring during May to July is the leaf curl. The curling of the leaves occur due to thrips belonging to the order Thysanoptera of class Insecta and species *Thrips tobcai.* The damage to leaf is done by adults as well as the nymphs, which commences feeding on plant sap by lacerating the leaf tissues. The tissue around the feeding punctures become dry and turns brown giving the leaves a flecked appearance. Ultimately, the margin of the leaf gets curled with secondary and tertiary curling and the leaf becomes needle like with wavy margins. This causes drastic reduction in the leaf area and also induces premature leaf fall. So it creates an obstacle for the growing stages of the larvae as a nutritional deficiency.

Control

Spraying of phosphomidon (0.03%) twice at an interval of 15 days during the month of April-May has been found to be effective in checking the spread of the disease. Moreover, the plantation area (Plot) which is affected during summer should be ploughed deeply followed by fumigation in winter in order to reduce the hibernating larvae of the thrip species (Table 3.1).

3.4. Seed Technology

Quality production of seed is the pivotal factor for a better stand of sericulture practice. Production and supply of quality seed in required quantity is one of the prerequisites of sericulture industry. The establishment of healthy silk insect eggs for production is known as *grainage* (French: *Graine*-seed). The selection and preservation of seed cocoons, the preparation of disease free layings (Dfls) and their disinfection and incubation are among the vital aspects of grainage. In tasar sector, the seed production still to a large extent is done by the rearers themselves. In the absence of proper seed organisation for the supply of quality tasar silk insect seed, the rearers have to depend on their own resources and the forests. The seed cocoons, thus, raised or procured by them are used for the preparation of layings without microscopic examination of the mother moths, proper

Table 3.1 : Control against pests, diseases and deficiencies of tasar food plants

Pest/disease	*Damage*	*Control*
Stem borer (***Sphenoptera konbierensis)***	Larvae bore into Arjun and Asan shoots	Plugging with cotton soaked in Petrol/Kerosene/EDCT and pasting with mud. Light traps for adults
Gall fly ***(Phylloplecta hirsuta*** and ***Trioza*** sp.)	Young flies produce ugly galls on Arjun & Asan foliage	Dimecron (0.03%) or Rogor (0.5%) spray
Termite ***(Microtermes, Odontotermes*** and ***Trinervitermes*** sp.)	Attack food plants at all stages of plant growth	Termitarium should be un-earthed by pouring crude oil emulsion/ aldrin (30% E.C.) emulsion. Aldrin dust (5%) mixed with soil (20 kg/ha)
Beetle ***(Anomala*** sp.)	Feeds on leaves of host plants	Tillage operation and aldrin dust (5%) at 20 kg/ha.
Stem - canker	Tumor on Asan twigs which bursts into ulcers	Heavy pruning and burning of infested twigs, dressing of cut ends with Bordeaux paste after pruning
Leaf curl	Crinkling of leaf surface in Asan and Arjun	Pruning, spraying with Zincop (0.3%)
Root rot ***(Gauoderma lucidum)***	Roots of Asan and Arjun decompose and die	Burning of dead plants. Digging of isolation trenches (1.5x0.7x0.3)m to isolate healthy from infected tracts
Molybdenum deficiency	Marked malformation of leaves and poor head formation	Liming of acid soils and application of ammonium-molybdate at 1.5 to 2.0 kg/ha.
Zinc deficiency	Green lateral leaf veins and mottling	Spraying with Zincop (0.3%).

disinfection and incubation of eggs. The unexamined seeds, thus, possess risk of multiplication and resulting in spread of disease and uncertain yield of cocoons. Moreover, the wild nature of the tropical tasar silk insect is reflected in its nonuniform and erratic emergence, coupling, egg laying and also some other habits. These are the main setback due to unsystematic handling and unscientific approach of grainage operation. For the preparation of quality seed, following steps may be undertaken: (i) systematic seed selection, (ii) procurement of seed cocoons, (iii) transportation, (iv) preservation, (v) egg preparation, (vi) moth examination, (vii) egg incubation, (viii) emergence (ix) coupling, (x) egg laying and (xi) hatching.

3.4.1. Systematic Seed Selection

For seed, the healthy, well-formed and tough cocoons are visually and tactually selected after harvesting and the dead, deformed and otherwise unsuitable cocoons are discarded. Currently each state has a programme of production and supply of certain quantity of commercial tasar silk insect seed to the rearers. This is done by purchasing the seed cocoons and organising its grainage operation. The quantity produced unscientifically is much less than the quantity required. The situation, thus, warrants that the rearers purchase seed cocoons and produce their own seed. Obviously, such seed remains unexamined. In order to streamline the system a four tier organisation was evolved by Central Tasar Research and Training Institute (CTR & TI), Ranchi in collaboration with state Government during 1978-79. Under this system the Research Institute maintains the silk insect seed stock and supplies to the nucleus or elite seed to its parental seed station and then supplies to different Basic Seed Multiplication and Training Centres (BSM & TCs) which further produce the basic seed and supply to the Pilot Project Centres (PPCs) of the states. So by this it has now been possible to select good quality and disease free seeds.

3.4.2. Procurement of Seed Cocoons

In order to lay hands on healthy and suitable seed cocoons, certain vital background information like fecundity, hatching, effective yield and disease incidence are collected. However, if such informations are not available, the cocoons to be procured should be uniform in size, shape and colour which roughly reflect the quality of seed cocoons. Hard and stiff cocoons are invariably better in this respect. Sample testing is carried out with the pupae for pebrine disease. The higher frequency or live pupae ensure better recovery of layings. The seed cocoons, thus, procured should have at least 5-10 days for emergence.

3.4.3. Transportation of Seed Cocoons

Transportation of seed cocoons from one place to another is inevitable for obvious reasons. However, this step is a vital one in the preparation of eggs and therefore, it is required to be attended to with utmost care considering its transportation from far off places. During transportation, the seed cocoons are packed loosely in the bamboo baskets or synthetic baskets or any perforated boxes in small quantities. Transportation is effected during the cooler hours of the day or in the mid night to avoid damage due to high temperature. So the seed cocoons are invariably protected from direct sun light. On consignment, the cocoons are unpacked and hung in the grainage room in the form of garland. Cocoons are transported during diapause preferably in December and January.

3.4.4. Preservation of Seed Cocoons

Immediately after the arrival of cocoons, they are spread in a single layer on a clean infection free floor. Screening is conducted to eliminate the defective cocoons such as flimsy, uzifly infected, deformed, melted and dry ones by using gloves on hands. The sorted cocoons are tied in the form of garland with the help of peduncles. The garland, thus made are hung with some support from the top leaving some space for proper aeration and operational movement. Suitable measures to overcome the damage caused by rats are also taken. The cocoons, harvested out of second and third crops in case of bivoltine and trivoltine respectively, enter into diapause extends from last part of August to last part of May and in bivoltine and trivoltines from November to last of May and from January to last of May respectively (Fig.3.29). During these periods, the pupae experience extreme temperature and relative humidity not only causing some mortality but also adversely affecting the vigour of ensuing progenies. The optimum temperature and relative humidity for seed cocoon preservation is 25°C and 45-48 per cent respectively. However, the mortality may be reduced considerably by preserving the cocoons in a mud walled model thatched grainage house where fluctuation of temperature is quite low. The condition further improves in an underground place where fluctuation is minimal.

Extra care during summer months has to be taken due to temperature interference. This may be achieved by putting false ceiling of bamboo mats and providing sand beds on the floor sprinkling water over it from time to time. Use of wet gunny cloths on doors and windows aided by small windows close to the ceiling have proved effective in reducing the heat radiation to a considerable extent. Besides, the doors and windows need

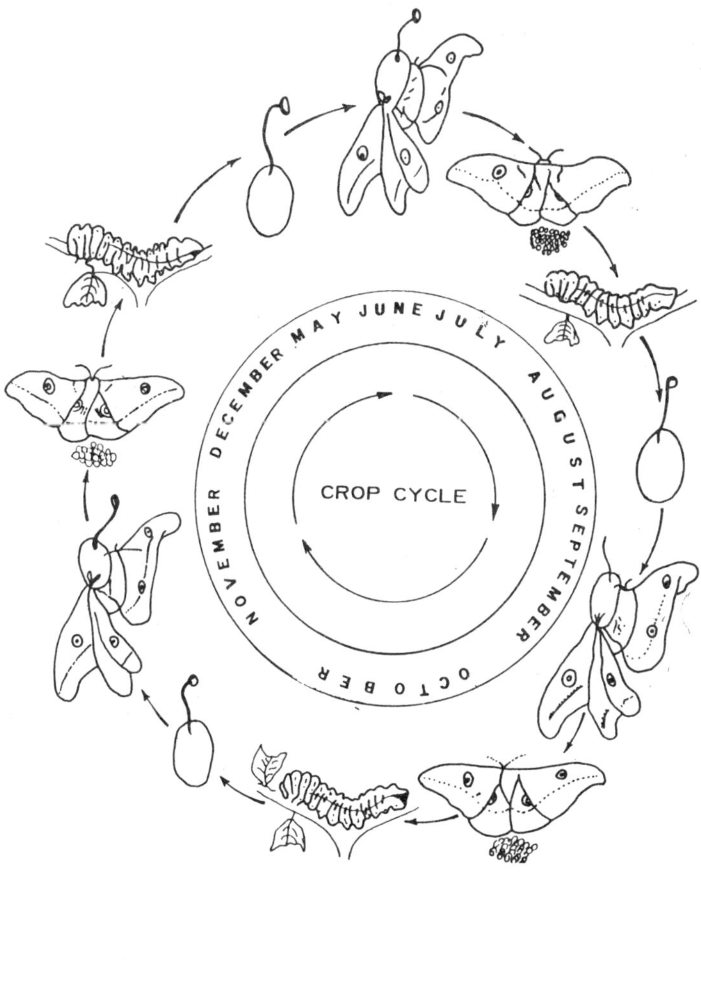

Fig. 3.29 : Life History of Tasar Moth (VOLTINISM)

to be kept open in summer months during the night and early hours of the day for proper ventilation. The seed cocoons once handled in a grainage, leave behind all sorts of contamination. Therefore, it becomes imperative to disinfect the grainage alongwith appliances thoroughly before fresh seed cocoons are consigned. The disinfection is carried out with 2% formalin spray followed by BHC 5% dusting. Fumigation of rooms for 72 hours has been found to be very effective. Use of flame guns in and around the grainage house has also been effective in eliminating the contamination.

3.4.5. Emergence of Moth

The regular emergence of moth from the diapausing pupae starts with the outset of monsoon excluding a few stray cases. The moths from the harvest of first crop (July) usually emerge during August-September while in case of trivoltine race, second crop harvest commences during October followed by further emergence during late October and early November for 3rd crop (Fig. 3.29,30). Temperature and humidity are two important factors guiding the course of emergence. It has been observed that 26°C-28°C temperature and 72% to 80% relative humidity aid in regular emergence of moths. Generally the moths start emerging late afternoon but the peak period is late hours of evening, that is 9 PM to 11 PM (Fig. 3.31). Emergence in the natural population takes a month, but with the cocoons from the first crop this period can be shortened and the emergence from those harvested on successive days can be synchronised for effective grainage by low temperature treatment. Most of the moths emerge from the peduncle portion, but very few also emerge from the posterior end of the shell, which is normally more in winter season (Fig. 3.32).

3.4.6 Mating of Moths

Mating or coupling takes place after 3 hours of emergence, the peak period being from midnight to 2 AM (Fig. 3.33). Moths generally prefer darkness and a temperature range of 26°C-28°C with 72%-80% relative humidity for effective coupling. Before allowing the female moths for oviposition, they are gently pressed to induce urination. Usually the tip of the wings of the female moths is cut so as to identify the mated female moths

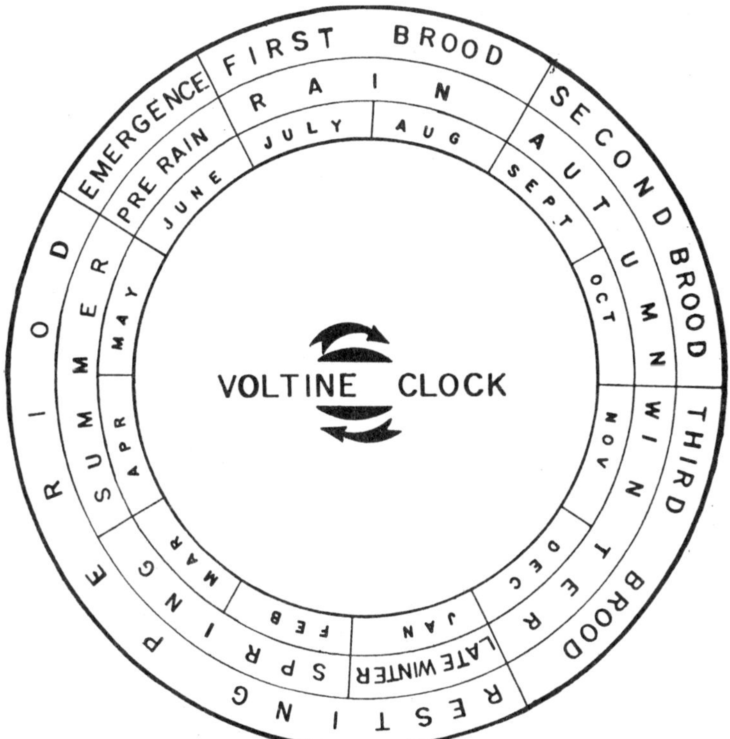

Fig. 3.30 : Voltine Clock of Tasar Moth

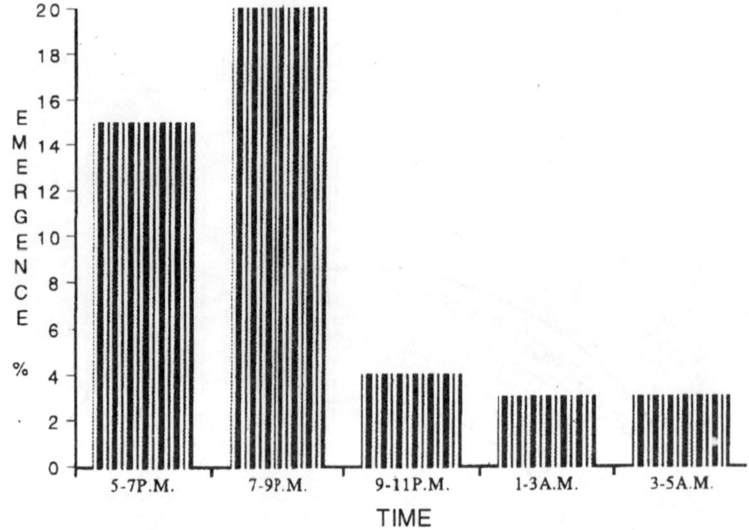

Fig. 3.31 : Emergence Percentage at Different Hours

conveniently. During the first crop when temperature remains high and humidity low, moths are kept under mosquito nets or wire mesh cages in outdoor conditions particularly during night hour for maximising the couplings. Usually males are used for single mating. However, they can be used for the second time too in case of shortage of male moths. In order to meet the shortage of male moths, the used ones are preserved at 5°C. Before they are brought to use for the 2nd time the male moths are exposed to the room temperature.

Coupling may be carried out in captivity using bamboo baskets, coupling cage or manually and in outdoor condition also. The outdoor coupling is more successful as compared to coupling in captivity (Fig. 3.34). In this outdoor mating, the females are placed on the bushes during early evening hour and then the males are released. The coupled pairs are collected early in the morning to avoid losses due to bird attacks. Although coupling continues for 10-12 hours, the duration of 3-4 hours is enough to secure complete fertilisation (Fig. 3.35). The duration of coupling has no direct relationship to the preoviposition period.

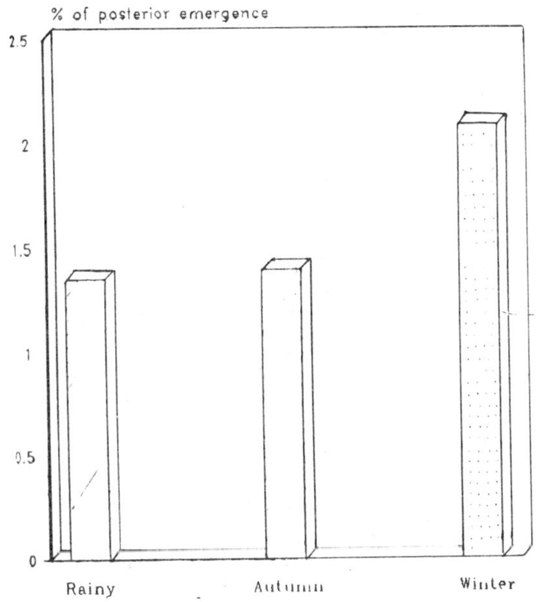

Fig. 3.32 : Posterior Emergence of Tasar Moth

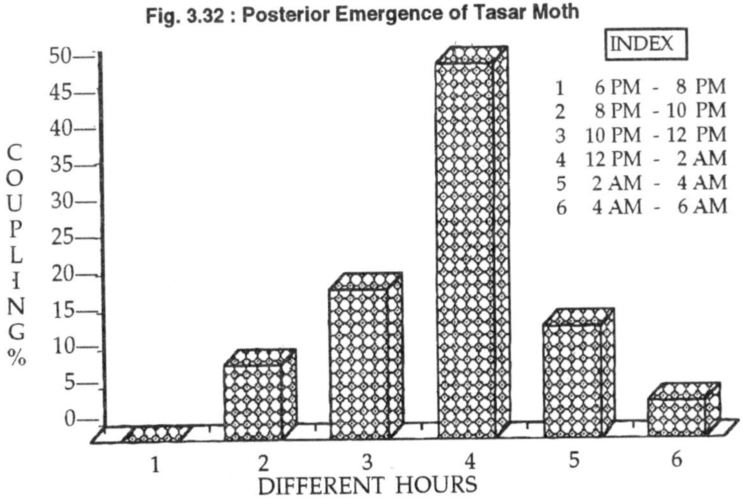

Fig. 3.33 : Coupling Aptitude at Different Hours

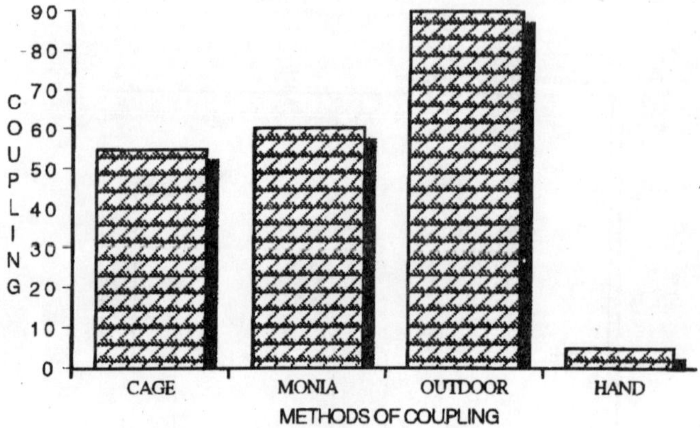

Fig. 3.34 : Methods of Coupling and its Percentage

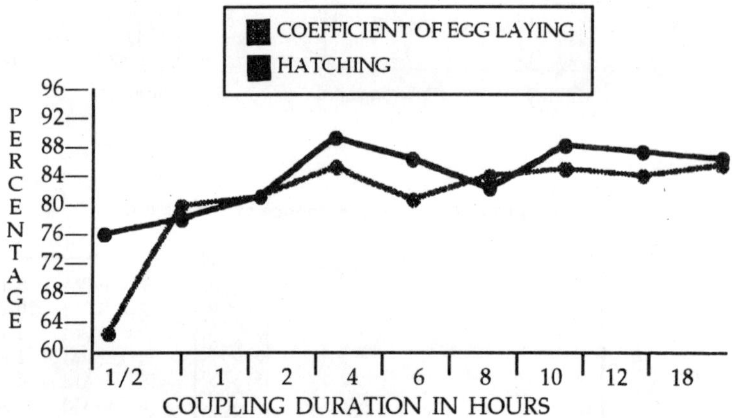

Fig. 3.35 : Effect of Coupling Duration on Oviposition and Hatching

3.4.7. Egg Laying

The wings of the mated female moths are clipped and they are kept in specially made earthen cups or card board cover or sweet packets or plastic boxes. Though the egg laying continues up to 7 days, the eggs up to 3 days are only collected for avoiding wide disparity during rearing. The eggs are deposited in batches of 5-10. Moths usually prefer darkness for egg laying. In

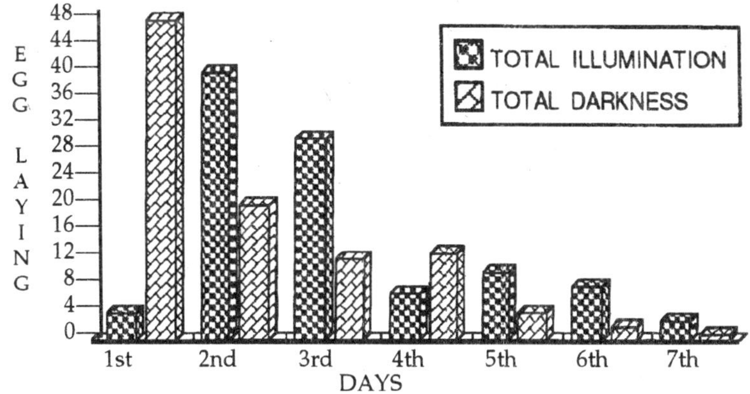

Fig. 3.36 : Effect of Photoperiod on Egg Laying

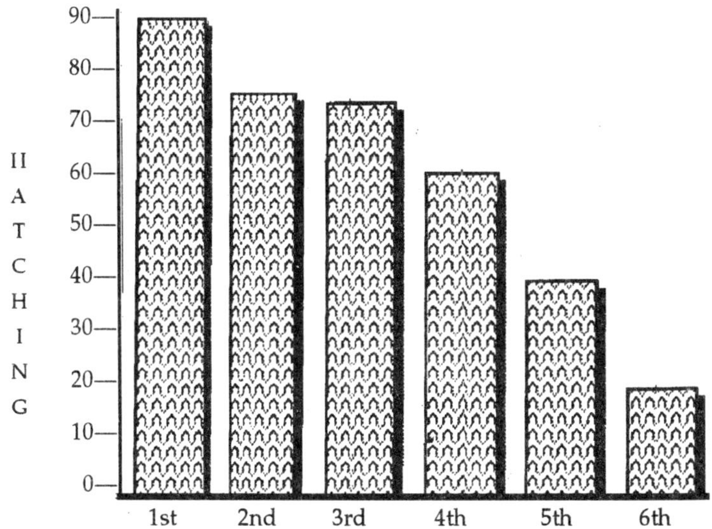

Fig. 3.37 : Hatching Behaviour of Eggs Laid on Different Days

complete darkness more than 50% of the eggs are laid on the first day whereas during total illumination laying is delayed (Fig.3.36). Most of the eggs (70-75%) are laid on the very first day (Fig. 3.37). In commercial grainage five moths are kept together for egg laying. This, however, is risky as the pebrine disease once revealed in any of the 5 moths, the entire lot of eggs need to be rejected. Therefore, the wastage get reduced when individual moths are kept separately for egg laying and are examined for the pebrine. A gummy substance discharged by the female moth helps the eggs in sticking to the egg laying surface. The earthen cups or any container used for collecting eggs are thoroughly disinfected before use.

3.4.8. Moth Examination

The moths after emergence from cocoon and mating, the female starts laying eggs. But pebrine infection in mother is very harmful for the next crop. Of all the diseases, pebrine is directly transmitted from the mother moth to the offspring through its eggs. Therefore, its spread is effectively checked with the examination of the mother moth. After the 3rd day of oviposition when majority of the eggs are laid, female moths are subjected to microscopic examination. For this purpose the tip of the abdomen is cut and crushed in a mortar and pestle. A drop of the suspension is taken for microscopic observation. The pebrine spores are identified by their bright oval body only under the microscope. Besides pebrine spores, the moths are also examined for viral and bacterial diseases. For viral disease, polyhedral inclusion bodies are looked for whereas white bacillus or cocci indicate bacterial disease. The eggs laid by diseased moths are rejected to prevent further spread of the disease and the eggs of healthy moths which are termed as disease free layings (Dfls) or certified eggs are retained for raising crop.

3.4.9. Egg Surface Sterilisation

After examination of moth, the eggs from healthy mother moth are collected in the cloth bags and washed with soap solution to remove muconieum. Thereafter, the eggs are disinfected with 5 per cent formalin solution for 5 minutes. The disinfected eggs are subjected to washing with distilled water and then dried in shade by spreading in single layer of blotting paper. The dried eggs are kept in laying boxes for incubation. Only eggs laid by healthy moth in sufficient number are passed for rearing. Moths which die in course of oviposition or lay eggs in less number are rejected and such eggs are not considered for rearing.

3.4.10. Packing and Transportation of Eggs

For transportation, the eggs are kept in small bags of muslin cloth of 9" x 4" size which can conveniently accommodate 100 Dfls. These bags are put in perforated plastic or bamboo baskets and carried as such. Sufficient space for aeration is given in between the bags. Under no condition the eggs are exposed to sun or high temperature as it affects the development of the eggs adversely. Thus, the eggs are transported within 6 days of oviposition to avoid hatching during transportation to distant places.

3.4.11. Incubation of Eggs

The egg boxes are placed in a room for incubation. In order to have better growth and development of embryo and uniform hatching, eggs are incubated at 28°C-30°C and 72 per cent-80 per cent relative humidity. By this operational convenience, it ensures more complete hatching successfully.

3.4.12. Hatching

Morning is the favourable time for hatching. The hatching of unincubated eggs begins on the tenth day and continues for six days. The hatching of properly incubated eggs starts on the 7th day or 8th day and is completed within two to three days. The percentage of hatching is higher in the eggs laid on earlier days (Fig. 3.37).

3.4.13. Refrigeration of Eggs

The eggs of *A. paphia* usually require seven days of incubation at room temperature and hatch on the last part of the 7th day or on the eighth day after oviposition. Refrigeration is a common means to delay hatching of other silk insect eggs.

One and two days of refrigeration of first (fresh or 0 day old) and second day (1 day old) eggs increases the incubation period to 12 days, 5 days more than the controlled ones. Three and four days of treatment to such eggs increases the incubation period to 13 days, days more than the control. One and two days of refrigeration increases the incubation period by two days beyond the control in third (2 days old) and fifth day (4 days old) and by one day in sixth (5 days old) and seventh

(6 days old) eggs. Similarly three and four days of refrigeration increases the incubation period by four days in third and fifth day eggs by three days in fourth and sixth days eggs and by two days in seventh day eggs.

The two-way refrigeration treatment differs significantly from the three day. Thus, the fresh and one day old eggs of *Antheraea paphia* refrigerated for three and four day show maximum increase of the incubated period amounting to six days.

4
Ecology

4.1. Geography and Topography of Tasar Producing Areas in Orissa

Orissa is one of the 25 states of Indian union and ranks third among tasar producing states (Fig. 4.1). It is situated within 81°-27′E to 87°-29′E longitude and 17°-49′N to 22°-35′N latitude and is divided into 30 districts. Physiographically it is divided into four natural blocks (Table. 4.1). The terrain ranges from 0

Table 4.1 : Physiography of Orissa

Sl.No.	*Principal Blocks*	*Districts*	*Topography*
1.	Northern Plateau Keonjhar Sundargarh Dhenkanal	Mayurbhanj	It is the part of Chhotanagpur Plateau of Bihar consisting of undulating uplands frequently intercepted by hill ranges. This is the major seat of tasar.
2.	Eastern Hills	Koraput Kalahandi Phulbani	It is the continuation of Eastern Ghat hill ranges. Hill ranges and foot hills harbour the tasar fauna.
3.	Central River Basin	Bolangir Sambalpur Dhenkanal	It consists of plain lands and river basin alongwith dense forest and hillocks.
4.	Coastal Plain	Puri Cuttack Balasore Ganjam	It consists of coastal plain lands and river basins in the East and hillocks and thin forest having tasar flora and fauna.

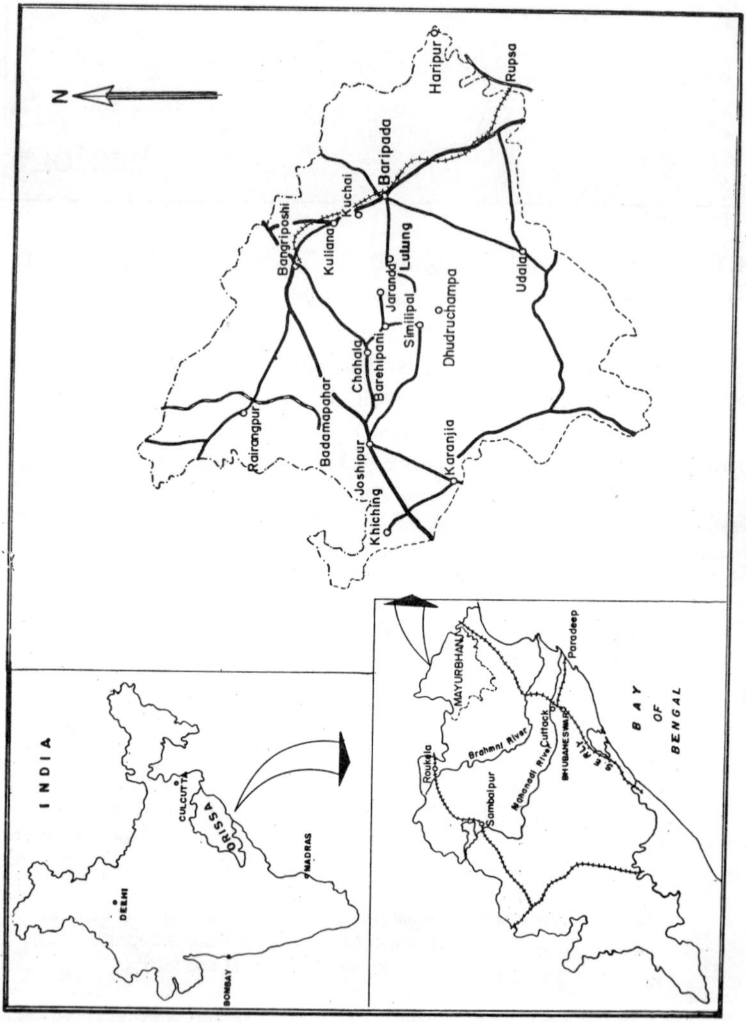

Fig. 4.1 : Location Map of the Study Site

to 1,500 metre above sea level (m ASL) having large areas of plain lands, hills, valleys, uplands, mountains, streams and rivers. Variations of topography, minerals, soil types, flora and fauna alongwith variation of principal tasar food plants are observed in an altitudinal gradient (Table 4.2).

Table 4.2 : Topography and tasar food plants of Orissa in altitudinal gradient

Sl.No.	*Altitude (m ASL)*	*Topography*	*Dominant Tasar Foodplants**	
			Common name	*Scientific name*
1.	50-300	Plain forests	Arjun Asan Sal	*Terminalia arjuna* *T. alata* *Shorea robusta*
2.	301-600	Hillocks and Valley forest	Asan Arjun Sal	*T. alata* *T. arjuna* *S. robusta*
3.	601-1,000	Hill forest	Sal Asan Arjun	*S. robusta* *T. alata* *T. arjuna*

*Numbered in order of decreasing density of food plants.

Although tasar cultivation is practised in almost all the districts of Orissa (Table 4.3), but the undivided district of Mayurbhanj occupies a prime position. Varieties of tasar flora, fauna, altitudinal range and ecological niche found in different districts of Orissa are all assembled in this district. Mayurbhanj ranks first in the quantity of tasar cocoon production, number of rearers and number of food plants. So the district Mayurbhanj represents a model and ideal place of tasar flora and fauna of the whole state.

Mayurbhanj is the land locked district of 10,418 sq km between 85°-40' to 87°-11' longitude East and 21°-16' to 22°-34' latitude North. Out of the total 3,790 sq km of tropical forest in the district, 2,750 sq km constitute the centrally located Simlipal hill forest, which divides the district into three natural divisions, that is (i) Eastern division, (ii) South-Western division, and (iii) North-Eastern division (Sinha 1981). The Eastern division slopes gently from the foot hills towards the sea and the two other divisions are mainly undulating planes rising and falling in gentle

Table 4.3: Districtwise tasar fauna and flora

Sl. No.	District	Altitude (m ASL)	TASAR FAUNA			TASAR FLORA		Distribution
			Species	*Occurrence*	*Frequency of Life Cycle*	*Food Plant*	*Occurrence*	
1.	Balasore	0-250	*A. mylitta*	Cultivated	3	*T. tomentosa* *T. arjuna* *S. robusta*	Wild Implanted Wild	1 Block
2.	Bolangir	350-600	*A. paphia*	Wild	2	*S. robusta* *T.tomentosa*	Wild Wild	2 Blocks
3.	Cuttack	0-350	*A. mylitta*	Cultivated	3	*T. tomentosa* *T. arjuna*	Wild & Implanted	1 Block
4.	Dhenkanal	150-1,150	*A. mylitta* *A. paphia*	Cultivated Wild	2 1	*S. robusta* *T. tomentosa* *T. arjuna*	Wild Wild Cultivated	6 Blocks
5.	Ganjam	0-850	*A. paphia*	Wild	1	*S. robusta*	Wild	1 Sub-division
6.	Kalahandi	400-900	A. *paphia*	Wild	1	*S. robusta* *T. arjuna*	Wild Cultivated	2 Blocks
7.	Keonjhar	350-900	*A. mylitta* *A. paphia*	Cultivated Wild	2	*S. robusta* *T. tomentosa* *T. arjuna*	Wild Wild Cultivated	Throughout
8.	Koraput	250-1,500	*A. mylitta*	Cultivated	2	*T. arjuna*	Cultivated	1 Block
9.	Mayurbhanj	50-1,500	*A. mylitta* *A. paphia*	Cultivated Wild & Cultivated	3 2	*S. robusta* *T. arjuna* *T. tomentosa*	Wild Cultivated Wild	Throughout
10.	Phulbani	350-1,250	*A. paphia*	Wild	2	*S. robusta* *T. arjuna*	Wild Implanted	2 Blocks
11.	Puri	0-350	—	—	—	*S. robusta*	Wild	Negligible
12.	Sambalpur	400-800	*A. mylitta*	Cultivated	2	T. tomentosa *T. arjuna*	Wild Cultivated	3 Blocks
13.	Sundargarh	400-1,000	*A. paphia* *A. mylitta*	Wild Cultivated	2	*S. robusta* *T. tomentosa* *T. arjuna*	Wild Wild Cultivated	Throughout

slopes. The terrain ranges from 50-1000 m ASL having a large number of hills, valleys and streams. The thickness of the forest and tallness of trees increase with the increase in the altitude of the place. At 50-300 m ASL, plane land and thin forest without hills are found. At 301-600 m ASL, medium forest with two low height scattered hills and at 601-1,000 m ASL, very thick forests and a number of ranges of high hills and mountains are found. Places above 1,000 m ASL mainly constitute hill peaks and are not easily approachable.

Various parameters such as the geography, topography, ecology, tasar flora and the frequency of life cycle of tasar silk moth in an altitudinal gradient including the places at the interval of 100 m ASL elevations from 50 to 1,000 m ASL are important for successful tasar culture. Below 50 m ASL, there is no rearing fields and above 1,000 m ASL is a hill peak which is absolutely unapproachable. At each study site, one rearing field, one rearing hut and one grainage house are provided. The rearing fields are demarcated in natural forests having adequate number of food plants for tasar silk moth rearing. Temporary rearing huts are constructed at each rearing field for keeping rearing appliances and for shelter and rest. Temporary grainage houses are either constructed or procured on a hire basis for storing live cocoons and observing their reproduction and preparing certified eggs for subsequent generation whenever required. Besides these infrastructure, a laboratory of the state Government is also set up for laboratory investigation and analyses.

4.2. Climatological Parameters

Different climatological parameters in an altitudinal gradient in all the cultivated sites play a significant role (Table 4.4). Environmental parameters like temperature, humidity, rainfall, duration of sun shine and day length are the regulating factors of the cocoon production. The observatories are installed at well exposed level spots of 54 metres (N - S) by 36 metres (E - W) away from high hills, buildings and water channels (Venkiteswaran 1956). But at higher altitudes, the presence of the tall trees, dense forests and high hills may not be avoided at all places due to the very topography of the place. A maximum and minimum thermometer placed inside the **Stevenson screen** at 1.3 metres above the ground aids in measuring the air temperature and hygrometer exposed in a Stevenson screen for humidity. The rain gauge erected on (0.6 x 0.6 x 0.6) m concrete foundation measures the rainfall. The sunshine recorder erected on a solid masonry pillar at a height of 3 metres above the ground records the duration of sunshine and the minimum radiation value for sunshine recorded is about 0.2 - 0.4 cal/cm^2/min. The day length is essential to note from the timing of sunset and sunrise which are related to the visibility of the upperbim of the sun on the horizon. Considering the life cycle pattern of tasar silk moth in nature and topography and dominant food plants of various places (Table 4.3), clubbing is made into three groups in an altitudinal gradient, that is one group for 50-300 m ASL, another group for 301 - 600 m ASL and a third group for 601-1,000 m ASL.

Table 4.4 : Geographical position and infrastructural facility in various zones

Sl. No.	*Study site*	*Altitude (m ASL)*	*Longitude (East)*	*Latitude (North)*	*Infrastructure*
1	Durgapur	50-100	86°—55'	21°—45'	State Govt. Research Farm
2	Kaptipada	101-200	86°—35'	21°—30'	Govt. Tasar Seed Station
3	Bangriposi	201-300	86°—33'	22°—10'	Govt. Tasar Seed Station
4.	Bisoi	301-400	86°—27'	22°—11'	Co-operative Rearing Unit
5.	Thakurmunda	401-500	86°—10'	21°—32'	Govt. Seed Station
6.	Kendumundi	501-600	86°—08'	21°—40'	Natural Rearing Field
7	Bhanjabasa	601-700	86°—25'	21°—35'	Natural Rearing Field
8.	Chahala	701-800	86°—25'	21°—58'	Meteorological Unit
9	Nigirda	801-900	86°—10'	21°—52'	Tasar food plants & Tribal Village
10	Nawna	901-1,000	86°—28'	21°—50'	Tasar food plants, Tribal village and Meteorological Unit
11	Meghasana	1,001-1,500	86°—25'	21°—40'	Hill peak unapproachable

Corroborating with different events of the life cycle of tasar silk moth that is emergence, rearing, pupation and voltinism with that of the season and environmental parameters, the monthly observation are also grouped into different seasons. Depending upon the frequency of life cycle of the race, the reproduction coincides with rain (July-August), autumn (September-October) or winter (November-December) seasons. The silk-moth remains under pupation for a minimum period of six months from January to June when extreme coldness (Late winter - January) and extreme hotness (summer - April and May) is experienced intercepted by spring (February and March). All the resting pupae emerge during pre rain (June) period.

4.3. Host Plant, Arboriculture and Topography

Tasar silk insects are polyphagous in nature. According to the preference of food plants, certain plants yield a better crop of cocoons. These are referred to as *primary food plants* but more than two dozens of food plants of secondary importance are also known (Table 4.5). Of these, tasar insect thrives well on three main food plants : Asan *(Terminalia alata)*, Arjun *(T. arjuna)* both of the family Combretaceae and Sal *(Shorea robusta)* of the family Dipterocarpaceae. The food plants of tropical silk-moth grow luxuriantly at low altitudes (600 m ASL) extending mainly up to torrid zone ($23^1/_2$°N and $23^1/_2$°S latitude) and rather sparsely up to 40° latitude in either direction. Tropical tasar food plant is dominated by *Terminalia* and *Shorea* and in India Asan *(T. tomentosa)* is the most important food plant of tropical tasar silk. Recently Akai *et al.*(1990) have attempted the culture of tasar silk insect by introducing artificial diet. The food plants do not influence the colour of the cocoon or silk, but contribute to its quantitative characters.

Seedling propagation of tasar food plants, though slow, is a common practice as they cannot be easily propagated asexually. Tasar food plants as found in nature are tall trees irregularly distributed in forests or on bundhs of paddy fields. Because of the consequent difficulties of overseeing the larvae, mortality from pests and disease is high particularly during the early instars. As the foliage consumption of the larvae until the third instar is only 15-20 per cent, this problem is solved effectively by rearing them in limited systematic plantation. Economic plantations of closely spaced (1.2 m x 1.2m) Arjun *(T. arjuna)* of

Table 4.5 : Host plants of *A. paphia*

Sl. No.	*Common name*	*Scientific name*	*Family*
		Primary Food Plants	
1	Asan	*Terminalia alata* Heyne ex Roth	Combretaceae
2	Arjun	*Terminalia arjuna* (DC.) Wt. & Arn.	Combretaceae
3	Sal	*Shorea robusta* Gaertn. f.	Dipterocarpaceae
4	Sidha	*Lagerstroemia parviflora* Roxb.	Lythraceae
5	Jarul (Patoli)	*Lagerstroemia flosreginae* Retz.	Lythraceae
6	Saoni	*Lagerstroemia indica* Linn.	Lythraceae
7	Ber (Berkoli)	*Ziziphus mauritiana* Lam.	Rhamnaceae
8	Arjun	*Hardwickia binata* Roxb.	Caesalpiniaceae
		Secondary Food Plants	
1	Haritaki	*Terminalia chebula* Retz.	Combretaceae
2	Bahera	*Terminalia bellirica* (Gaertn.) Roxb.	Combretaceae
3	Jangli Badam	*Terminalia catappa* L.	Combretaceae
4	Kinjal	*Terminalia paniculata* Roth	Combretaceae
5	Asan variety	*Terminalia alata* Heyne ex Roth	Combretaceae
6	Asan variety	*Terminalia glabra* Retz.	Combretaceae
7	Ber	*Ziziphus mauritiana* Lam.	Rhamnaceae
8	Ghonto	*Ziziphus xylopyrus* (Retz.) Willd.	Rhamnaceae
9	Kantaikoli	*Ziziphus rugosa* Lamk.	Rhamnaceae
10	Dhaura	*Anogeissus latifolia* (Roxb,exDC.) Wall. ex-Guill & Perr.	Combretaceae
11	Jamun	*Syzygium cumini* (L.) Skeels	Myrtaceae
12	Kumbi	*Careya arborea* Roxb.	Lecythidaceae
13	Sal variety	*Shorea roxburghii* G. Don	Dipterocarpaceae
14	Teak	*Tectona grandis* L.f.	Verbenaceae
15	Aswatha	*Ficus religiosa* L.	Moraceae
16	Jari	*Ficus retusa* L.	Moraceae
17	Jari	*Ficus tseila* Roxb. ex. Buch.-Ham.	Moraceae
18	Jari	*Ficus bengamina* Linn.	Moraceae
19	Kanchana	*Bauhinia variegata* Linn.	Caesalpiniaceae
20	Simuli	*Bombax malabaricum* DC.	Malvaceae
21	Simuli variety	*Bombax heptaphyllum* DC.	Malvaceae
22	Gongai	*Melastoma malabathricum* L.	Melastomataceae
23	Khirikoli	*Carissa carandas* Linn.	Apocynaceae
24	Mahula	*Madhuca indica* J. Gmelin	Sapotaceae
25	Rai	*Rhizophora candelaria* DC.	Rhizophoraceae
26	Nalbali	*Cipadessa baccifera* (Roxb.) Miq.	Meliaceae
27	Kumar chikni	*Canthium dicoccum* (Gaertn.) Teijsm. & Binnend	Rubiaceae
28	Dodonea	*Dodonaea angustifolia* L.f.	Sapindaceae
29	Cassia	*Cassia senna* L.	Caesalpiniaceae

2 to 3 years old saplings are suitable for rearing. The higher density of trees per unit area yields over 17 metric tons of foliage per hectare. Furthermore, being the medium bushy type with overlapping twigs, tasar food plants may be rain fed and grow in submarginal soils. Sal *(Shorea)* trees consists of half a dozen of better known species like S. *talura*, S. *vulgaris*, S. *cochinchinensis*, S. *thorelli*, S. *obtusa*, S. *hypochlora* and S. *robusta*. Of these *S.robusta* is abundant throughout Orissa and India (Table 4.6) excepting few places and provides a rich base for nature grown tasar cocoons. Among the secondary food plants like Ber or Berkoli *(Ziziphus* sp.*)* is very important which is indigenous to India, Burma, Malaysia and Bangladesh. The characteristic feature of Arjun plant *(T. arjuna)* is quick growing for which cultivation is mostly preferred by rearing these saplings. The effective rate of rearing (ERR) depends on bush type alongwith the nature of sprouting. Natural sprouting makes the leaves suitable for first rearing. Plucking of consumed leaves and light pruning after first rearing help in getting new flush which is fit for rearing in about 4-6 weeks. Moreover, raising bushy type of plantation with closer spacing indicates that with the decrease in spacing, the yield of foliage per plant goes down whereas the yield per unit area decreases. Therefore, plantation with four feet (4') spacing for tasar culture supports 600-650 egg layings thereby increasing the rearers' income substantially. In India about 12.43 million hectares of unexploited tasar food plants in the tropical and temperate belts could be put to use without investment. The unutilised resource exploitation would not only offer rural population an attractive source of income but also permit simultaneously conservation and economic exploitation of forests.

Food plants are one of the major determinants of the fate of the crop. Waterlogging in the fields during rains raises the relative humidity, excreta and dead larvae which fall into water, gradually putrefy and pollute the environment. So it is necessary to avoid low lying areas. Thick patches of food plants of 20-30m tall with rich foliage and well spread crown are ideal. For these ideal conditions, the pits (30 cm^3) are filled with the mixture of solid 10-20 kg compost and 100-200 gm each of *karanj* cake, bone-meal, wood-ash and aldrin dust (5 per cent). Transplanting is desirable in early monsoon (June-July) for economic and block plantations. Of the three common training systems - central leader, open centre and modified leader - the last is the most suitable for economic plantation in which the central axis is not allowed to grow indefinitely. In natural plantation, trees are pollard 1-2 m above the ground level in March and April every 3-4 years. After the plants are grown, these are protected from pests, diseases and mineral deficiencies.

Table 4.6 : Distribution of primary food plants of tropical tasar

Food Plant	*Distribution*
Terminalia sp.	India, Nepal, Burma, Laos, Khmer, Sri Lanka, Indonesia, Malaysia, New Guinea, Australia and Africa
T.tomentosa var.	North India up to Nepal and Central India
T. arjuna	Central and Eastern India
Shorea robusta	Throughout the tropics and subtropics
Lagerstroemia sp.	India, Sri Lanka, China, Vietnam, Malaysia, Indonesia, Australia
L. parviflora	Base of western Himalayas and south of Indo-Gangetic plain, Central India, Nepal, Burma, North-Eastern India.
L. speciosa	India, Vietnam, Malaysia, China and Sri Lanka
Hardwickia binata	Tropics particularly Africa and Western peninsular India
Ziziphus mauritiana	India, Afghanistan, Malaya, Archipelago, China and Australia

4.4. Conservation of Tasar Biodiversity and Sustainable Development

Everywhere, man lives in interspecific animal and plant communities and like any other organism, he depends on his environment for food, water, air, space, shelter and clothing. Man, being a dominant organism of most ecosystems, controls and modifies environments more extensively than any other organism. In fact, there have been significant changes in the natural environments due to man's intervention and his rapid progress in industrialization, urbanization, colonization, agriculture, mining, transportation and technology. Thus, deforestation has provided land for agriculture and rural inhabitation, rural land has been converted into urban-settlements and open spaces are fast vanishing. The mass-scale destruction of flora and fauna has become detrimental to ecological balance. Despite man's awareness to his ecological malpractices, still in many tropical lands deforestation is occurring at increasing rates and the people believe that their forests are unlimited.

Forest is a priceless wealth to the human society and an important renewable natural resource. The total forest area of the world in 1900 was nearly 7000 mha. By 1975 it was reduced to 2890 mha and if it continues like this by 2000 A.D. it would be merely 2370 mha. The major reduction will be in tropics and subtropics (40.2 per cent) being only 0.6 per cent in temperate areas. The tropic rain forest was originally 1000 mha in 1900, which was reduced to 940 mha by 1975. Maximum reduction has been estimated in India, Burma and Sri Lanka. The rate of destruction of forests is, thus, alarming in developing countries like India. Tropical forests occupy only 7 per cent of the land surface area and have a bewildering biological diversity. Tropical forests are considered as the lungs of the earth and hence have aptly been called as the life support system. Merciless clearing of forests is at the rate of 6 to 8 mha per year. By the end of 2000 A.D. about 10 per cent of the existing species would cause to exist, amounting to a loss of 25,000 plant species. The rapid loss of plant species (i.e. one species per day) termed as genetic erosion will not only wipe out the bulk of renewable plant resource, but there will also be a tremendous loss of vast reservoir of genetic diversity that could be a potential loss in future crop improvement programmes.

Silviculture is a branch of forestry which remains concerned with the establishment, development, care and reproduction of monocultures of valuable timber trees such as teak, sal, sheesham, kel, arjun etc. In India, we have only 0.97 per cent of the country's land area under National Parks, which corresponds to a mere 4.90 per cent of the forest area of the country. While we should have 33 per cent of the total land area of the country under forest cover for ecological reasons, we actually have only 21 per cent or so approximately now under forest cover. Even out of this inadequate forest area we have a large chunk of degraded forest area. And worse still is the fact that even now there is tremendous pressure for diversion of more forest areas for non-forest purposes, which has prompted Govt. of India to enact the forest conservation Act, 1980.

Conservation is the most efficient and most beneficial utilization of natural resources. It is chiefly concerned with the management of the natural resources of the earth, taking into consideration their proper use, preservation and protection from destructive influence and misuse. Precisely, conservation is the national use of the environment to provide a high quality of living for the mankind. The important objectives of conservation are :

(i) to maintain the essential ecological processes and the life support systems which have air, water, land, flora and fauna as the important elements.

(ii) to preserve "biological diversity" (i.e. both genetic and ecological diversity). Ecological diversity indicates the richness of the species, e.g. India is gifted by nature with about 45,000 species of plants and more than 65,000 species of animals.

(iii) to ensure "sustainable utilization" of species and ecosystem. Sustainable utilization means proper and planned utilization of natural resources so that a continuous yield of useful minerals, plants and animals may be obtained.

In view of the importance of forests and with a view to protect them, a "National Forest Policy" was announced by the Govt. of India in 1952. The policy has recommended that the forests be steadily increased to 33 per cent of the total area i.e. 60 per cent in the hills and 20 per cent in the plains. In fact, good forest cover is only 11 per cent and this too, is under heavy stress. The basic aims of the policy are :

(i) conservation of existing resources for safeguarding;

(ii) development and enlargement of the tree cover and resource-base to meet the basic needs of the people and the country;

(iii) to develop minor forest produce for providing sustenance for certain communities living near forests.

Thus, preservation of forests is even more important than conservation since forests are a priceless ecological resources, protecting land and water resources, controlling floods, checking wind erosion, storing and cycling nutrients and providing habitats for wildlife denizens. Finally, it constitutes a rich stock of valuable genetic resources, a common heritage of mankind.

Since utilisation of natural renewable resources is often linked with sustenance of man, the utilisation of such resources should be allowed for the "people in and around" the protected area, where the conservation of biodiversity is contemplated. With the unprecedented pressure of human population on such resources, it is just not possible to achieve the objective of the conservation of biodiversity. The unavoidable and sure fall out of such permission is overuse due to the rising demand on such resources, which probably for outstrips and sustainable limits of production—thus compounding the problem of degradation of the habitat or impoverishment of natural ecosystems, which ultimately effects adversely the natural poductivity also in a chain reaction. The process thus becomes counter productive. However, due to the global natural resource crunch, it has become transparent that we have no option but to conserve whatever biodiversity we are left with at present, mostly in the form of wildlife resources. The main aim of this conservation is to ensure perpetual propagation of these resources and may also include sustainable utilisation, as may be found possible. This will certainly call for *in situ* conservation and ensuring the natural process of evolution to continue without any hindrance. The loss of biodiversity will not only impoverish the gene pool resources, but will also adversely effect the natural productivity, the result of which will simply be dreadful.

Under this uncontrolled biotic interference and tremendous pressure on ecosystem, Govt. of India has decided to launch a special scheme called "ecodevelopment plan" to counter this dangerous trend. Social and economic development of a country is exclusively possible through conservation. As a developing agricultural country, India provides a lot of avenues through sericulture. For the greater interest of the country tasar biodiversity needs to be saved to nourish the various cross section of people. Realising the ecocidal alarming situation, tropical forests and tasar forests in particular should be preserved as "purifying lungs of the world" and "gardens of the biodiversity". The tremendous prodigality of nature has astonished and delighted humans and is ultimately the source of their sustenance. Yet, in using plants and animals for food and clothing, for building houses and for medicine, humans are endangering this immense richness of species. Therefore, "sustainable utilization" i.e. systematic, proper and planned use of the natural resources are essentially desirable: This lead to proper ecodevelopment by which ecological loss and the loss of the species could be arrested. On the other hand, human being as the largest predator group of the nature could yield the useful natural resources continuously for their development. So for sustainable development and equilibrium, sustainable utilization and ecodevelopment as two sides of a coin need to be strengthened.

References

Abe, H., Kabayashi, M. and Watanabe, H. (1990), Mosaic infection with a densonucleosis virus in the midgut epithelium of the silkworm *Bombyx mori*. J. Invert. Pathol,, **55**(1) : 112-117.

Abramova, I., YU, UL'masov, KH. A., Akopov, S.B., Karaev, K.K., Babaeva, A. KH. and Evgen'ev, M.B. (1991), Dynamics of synthesis of fibroin and heat shock proteins in different organs and at different stages of silkworm *Bombyx mori* ontogenesis. Prikl. Biokhim. Microbiol., **27**(1) : 147-156.

Ahsan, M.M. (1972), Exploitation of Sal for large scale rearing of tasar crop. Annu. Rept. Central Tasar Research, Ranchi, India, pp. 46-50.

Ahsan, M.M. and Griyaghey, U.P. (1972), Effect of interchange of food plants of tasarworm. Annu. Rept. Central Tasar Research, Ranchi, India, pp.43-46.

Ahsan, M.M., Jolly, M.S. and Khatri, R.K. (1974), Some preliminary studies on the various aspects of grainage in tasar silkworm A. *mylitta* D. Proc. 1st Int. Sem. Non-mulberry silks, Ranchi, India, pp. 158-159.

Ahsan, M.M., Srivastava, A.K. and Ansari, M.O.A. (1979), Preservation of Bogai cocoons at low temperature. Annu. Rept. Central Tasar Research, Ranchi, India, pp. 83.

Aizawa, K., Furuta, Y. and Nakamura, K. (1961), Selection of resistant strain to virus induction in the silkworm *B.mori*. J. Seric. Sci., **30**(5) : 405-412.

Akai, H., Suto, M., Nayak, B.K. and Asaoka, K. (1990), Rearing of Indian tasar by newly developed artificial diet. Proc. 1st Int. Conf. wild silkmoth, China, No.32, pp.25.

Alam, M.O. and Thangavelu, K. (1996), New trends in Bivoltine Tasar Silkworm rearing. Indian Silk, **35** (2 and 3) : 23-25.

Alamelu, M.G. and Matsuda, M. (1967), An improvement on the aspectic rearing of the eri silkworm *(P. cynthia ricini)*. Ind. J. Seric., **1** : 114-117.

Anantha Narayana, S.R. (1986), Silkworm rearing the present practices. Lectures on sericulture, Bangalore, India, pp. 51-59.

Arora, G.S. and Gupta, I.J. (1979), Taxonomic studies on some of the Indian non-mulberry silkmoths(Lepidoptera : Saturniidae : Saturniinae). Memoirs of the Zoological Survey of India, pp. 1-63.

Astaurov, B.L. (1940), Artificial parthenogenesis in the silkworm *B.mori* L. Acad. .Nawk., USSR, pp. 1-240.

Awaguchi, Y., Banno, Y., Koga, K. and Fujii, H. (1991), Effect of starvation stress on the growth and the haemolymph protein composition of the fifth instar larvae in *Bombyx mori.* Sci. Bull. Fac. Agric. Kyushu Univ. Japan, **45** (3-4) : 135-144.

Bahl, R.K., Pandey, R.K., Rao, P.R.T. and Sengupta, K. (1987), Induction of pupal diapause in oak tasar silkworm *Antheraea proylei.* Sericologia, **27** (3) : 567-570.

Barah, A., Phukan, S.N. and Samson, M.V. (1988), Variation in cocoon characters of *Antheraea assama* Ww. during different seasons. Sericologia, **28**(2) : 215-219.

Bharali, N. (1967), Som—the muga food plant : success in air layering. Indian Silk, **6** : 5-7.

Bharali, N. (1972), Studies on the density population of muga silkworm. A. *assamensis* Westwood. Ind. J. Seric., **11** : 1-5.

Bharali, N. (1974), Some studies on the ecology of muga (A. *assamensis* Westwood) culture in the north bank of upper Assam. Proc. Ist Int. Sem. Non-mulberry silks, Ranchi, India, pp. 67.

Bhaskar, M., Bharathi, D., Reddanna, P. and Govindappa, S. (1982), Growth pattern of silkworm larvae *B. mori* on exposure to prolactin. Ind. J. Seric., **21-22** : 33-35.

Blatter, E. (1929), The Indian species of *Terminalia* Linn. J. Ind. Bot. Soc., **8** : 245-262.

Boedts, B. (1990), The practices of tropical sericulture in Karnataka State, South India. Tropicultura, **8**(2) : 94-98.

Botenandt, A. and Karlson, P. (1954), Isolation of ecdysone from the pupae of *B. mori.* Z. Natureforsch, **9**B : 390-391.

Carne, P.B. (1966), Growth and food consumption during the larval stages of *Paropsis atomaria* (Coleoptera : Chysomelidae). Ent. Exp. and Appl., **9** : 105-112.

Census of India (1991), Directorate of Census Operations. Govt. of Orissa, pp. 1-4.

Chakravorty, R. and Das, B.C. (1979), Studies on diseases of tasar food plant and their control measures. Annu. Rept. Central Tasar Research, Ranchi, India, pp. 139-141.

Chakravorty, R. and Singh, U.N. (1983), Studies on sprouting behaviour of *Q. acutigenia* during different months. Annu. Rept. Central Tasar Research, Ranchi, India, pp. 374-379.

Chapman, R.F. (1982), The insects, structure and function. 3rd Ed. English Language Society and Hodder and Stoughton, pp. 474-476.

Chaudhury, A. and Sinha, A.K. (1994), Moulting behaviour of tropical tasar silkworm *Antheraea mylitta* Drury. Indian Silk, **33** (2) : 39-43.

Choe, K.J., Lee, D.H., Lee, W.Y., Lee, S.B., Jung, D.H. and Yang, S.Y. (1990), Survey on the structure and heating of silkworm rearing house in Korea. Res. Rep. Rural Dev. Adm., **32** (4) : 1-49.

Chowdhury, S.N. (1970a), Mugaculture. Directorate of Sericulture and Weaving, Govt. of Assam, pp. 15-76.

Chowdhury, S.N. (1970b), Ericulture. Directorate of Sericulture and Weaving, Govt. of Assam, pp. 11-40.

Chowdhury, S.N. (1974a), Ericulture. Directorate of Sericulture and Weaving, Govt. of Assam, pp. 42-45.

Chowdhury, S.N. (1974b), Mugaculture. Directorate of Sericulture and Weaving, Govt. of Assam, pp. 53-57.

Danilevskii, A.S. (1965), Photoperiodism and seasonal development of insects. Oliver and Boyd, London, pp. 1-283.

Dash, A.K. and Nayak, B.K. (1990), Studies on some oviposition behaviour of Godamodal ecorace of Indian wild tasar silkmoth *A. paphia* L. (Lepidoptera: Saturniidae). Proc. Ist. Int. Conf. wild silkmoth, China, No. 118, pp. 5.

Das, M.G., Jolly, M.S., Rao, M.R. and Singh, K.N. (1974), Economic plantations and scope of intercropping with *T. arjuna*. Proc. Ist. Int. Sem. Non-mulberry silks, Ranchi, India, p. 93-98.

Das, P.K., Sharan, S.K., Singh, R.N. and Sengupta, K. (1981), Studies on seasonal incidence of insect population on *T. arjuna* and *T. tomentosa*. Annu. Rept. Central Tasar Research, Ranchi, India, pp. 77.

Dawson, R.W. (1931), The problem of voltinism and dormancy in the polyphemus moth. J. Exp. Zool., **59** : 87-131.

Delvi, M.R. and Pandian, T.J. (1972), Rates of feeding and assimilation in the grasshopper *Poecilocerus pictus*. J. Insect Physiol., **18** : 1829-1843.

Deodikar, G.B., Chowdhury, S.N., Bhuyan, B.N. and Kshirasagar, K.K. (1962), Cytogenetic studies in Indian silkworms : Chromosome number in muga silkworm *A. assamensis* Ww. Curr. Sci., **31** : 147-148.

Devasurappa, L. (1986), Silk industry in Karnataka. Lectures on sericulture, Bangalore, pp. 32-34.

Devaiah, M.C., Rajasekhargouda, R., Subas, Y. and Govindan, R. (1985), Growth and silk production in *Samia cynthia ricini* Boisduval fed on 4 different host plants. Ind. J. Seric., **24** (1) : 33-35.

Dutta, R.K. (1988), Indian sericulture : Past, present and future. Int. Cong. Tropical sericulture practices, 18-23 Feb., pp. 217-240.

Evans, A.C. (1939), The utilization of food by certain lepidopterous larvae. Trans. R. Ent. Soc., London, **89** : 13-22.

FAO Agricultural services Bulletin (1987), Manuals on sericulture. Non-mulberry silks Vol. 4 and Mulberry cultivation Vol. 1. Central Silk Board, Bangalore, India, **1** : 1-97 and **4** : 1-278.

Figimoto, N. and Naomasa, K. (1961), Studies on the pigments of cocoons. J. Seric. Sci., **29** (5) : 420-424.

Ford, E.B. (1956), Moths. Collins, London, pp. 9.

Fracker, S.B. (1915), The classification of lepidopterous larva. Illinois Biol. Monogr., **2** : 1-169.

Fujimato, H., Sujimato, S. and Kawakami, K. (1958), Studies on the pigments of cocoons in the silkworm *B. mori.* J. Seric. Sci., **27** : 388-392.

Fukuda, S. (1953), Determination of voltinism in multivoltine silkworm. Proc. Jap. Acad., Tokyo, **29** (7) : 385-389.

Fukuda, T. (1967), The development of an artificial diet not containing leaf powder of any food plants for eri silkworm raising. J. Seric. Sci., **36** : 309-314.

Fukuda, T. and Higuchi, Y. (1963), Artificial food for oak silkworm raising. Agri. Biol. Chem., **27** : 99-192.

Gertsch, W.J. (1979), American spiders. Van Nostrand Reinhold Co., New York, pp. 52-65.

Ghosh, S.S. and Lamba, P.S. (1976), Studies on preservation of cooked Daba cocoons for reeling. Annu. Rept. Central Tasar Research, Ranchi, India, pp. 144-147.

Ghosh, S.S. and Sengupta, D. (1980), Studies on reeling of emerged tasar cocoons. Annu. Rept. Central Tasar Research, Ranchi, India, pp. 107-108.

Gopalchar, A.R.S. (1978), Sericulture industry in India; potentials and prospects. Central Silk Board, Bombay, pp. 1.

Gopinath, G.R. (1994), Drip irrigation for mulberry gardens. Indian Silk, **33** (2) : 27-31.

Govindan, R., Magadum, S.B., Bheemana, C. and Narayanswamy, T.K. (1987), Influence of mulberry varieties on cocoon weight, ovariole length, ovariole egg number and fecundity in silkworm *B. mori* L. Sericologia, **27** (1) : 25-34.

Govindan, R. and Narayanswami, T.K. (1985), Studies on double cocoons in silkworm *Bombyx mori* L. and their utilization in laying preparation. Entomon, **10** (1) : 59-62.

Green, T.L. (1932), Function of the spiracles of insects. Nature, **129** : 582.

Griyaghey, U.P., Jolly, M.S. and Kumar, P. (1976), Studies on the thermic control of microsporidiosis of the tropical tasar silkworm *A. mylitta* D. Ind. J. Seric., **15** : 27-30.

Gromadska, M. (1960), Observations on the thermal ecology of developing eggs of *A. pernyi.* Ekol. Polska Waram, **8** : 303-314.

Haga, A., Tsukada, M., Yasumara, S. and Aoki, H. (1991), Physical properties of silk sericin expected from cocoon shells of silkworm varieties with different values of sericin solubility. J. Seric. Sci., **60** (1) : 23-27.

Hamamura, Y. (1959), Food selection by sillkworm larvae. Nature, **183** : 1746-1747.

Hamamura, Y., Nayashiya, K., Naito, K., Matsura, K. and Nishida, J. (1962), Food selection by silkworm larvae. Nature, **194** : 754-756.

Hanumappa, H.G. (1986), Economics of sericulture, microperspectives. Lectures on sericulture, Bangalore, pp. 141-149.

Harade, H. (1921), Relation between moisture during incubation and voltinism in the bivoltine silkworm. Rep. Ueda. Seri. Coll., **4** : 125.

Hasegawa, K. and Ata, M. (1972), Penetration of phytoecdysone through the pupal cuticle of silkworm *B. mori.* J. Insect Physiol., **18** : 959-971.

Hayshiya, K. (1959), Studies on the pigments of cocoons. J. Seric. Sci., **28**(1) : 27-32 and **28** (2) : 84-87.

Horie, Y. (1962), Effect of various fractions of mulberry leaves on the feeding of the silkworm *Bombyx mori* L. J. Seric. Sci., **31** : 258-264.

Hyde, N. (1984), The queen of textiles. National Geographic, **165** (1) : 2-48.

Ibotombi, N., Bhaginath, Th. and Kundu, S.C. (1988), Pachytene synaptonemal complex analysis in Indian eri silkworm *Philosamia cynthia*. Sericologia, **28** (2) : 195-199.

Ichikawa, M. and Nishitsutsuji, J. (1952), Studies on insect metamorphosis. II. Determination of the critical period of pupation in eri silkworm *P. cynthia ricini*. Annot. Zool., Jap., **25** : 143-148.

Ikeda, M., Yaginuma, T. and Yamashita, O. (1991), Developmental profile and tissue localization of the protease responsible for vitellin degradation in the silkworm *Bombyx mori*. J. Seric. Sci., **60** (3) : 178-185.

Isa, Md. and Thangavelu, K. (1986), Sericigenous flora of North Eastern India. Lectures on sericulture, Bangalore, pp. 126-128.

Ishmaev, A.M. (1937), The morphology and post-embryonic growth of the silk secreting gland of *A. pernyi*. Rev. Zool. Russe., **16** : 239-246.

Ito, T. (1967), Nutritional requirements of the silkworm *Bombyx mori* L. Proc. Jap. Acad., **43** : 57-61.

Jayaswal, J. (1973), Studies on preservation of cocoons of *A. pernyi*. Annu. Rept. Central Tasar Research, Ranchi, India, pp. 231-233.

Jayaswal, J., Subramanyam, R.K., Patnaik, A.K. and Sengupta, K. (1981), Development of artificial diet for tasar silkworm. Annu. Rept. Central Tasar Research, Ranchi, India, pp. 59-60.

Jolly, M.S. (1972), A new technique of tasar silkworm rearing. Indian Silk, **11** : 5-8.

Jolly, M.S. (1974), New dimensions of tasar industry in India. Souvenir, Central Silk Board, Bombay, **XII** : 83-86

Jolly, M.S. (1976), Package of practices for tropical tasar culture. Central Tasar Research, Ranchi, India, pp. 32.

Jolly, M.S. and Sen, S.K. (1972), Infection of *A. mylita* D. (Lepidoptera : Saturniidae) by a microsporidian (*Nosema* sp.). Ind. J. Seric., **11** : 52-57.

Jolly, M.S. and Sen, S.K. (1974), Pattern of follicular imprints in egg shell - a species specific character in *Antheraea*. Ind. J. Seric., **13** : 36-43.

Jolly, M.S., Waratkar, V.B. and Prasad, S. (1967), A new mutant of tasar silkworm *A. mylitta* D. Ind. J. Ent., **29** : 389-390.

Jolly, M.S., Chaturvedi, S.N. and Prasad, S.A. (1968), A survey of tasar crops in India. Ind. J. Seric., **1** : 50-58.

Jolly, M.S., Sinha, S.S. and Sinha, A.K. (1970), Extraction of sex-pheromone from the moths of *B. mori* L. Science and Culture, **36** : 161-162.

Jolly, M.S., Sinha, B.R.R. Pd. and Sengupta, A.K. (1972), Chromosome number of *Antheraea frithii* Mr. Annu. Rept., Central Tasar Research, Ranchi, India, pp. 62-64.

Jolly, M.S., Ahsan, M.M. and Vishakarma, S.R. (1973), Rearing performance of tasar silkworms with interchange of food plants. Annu. Rept., Central Tasar Research, Ranchi, India, pp. 67-69.

Jolly, M.S., Das, M.G. and Rao, M.R. (1974), Raising of economic plantation of *T. arjuna*. Annu. Rept., Central Tasar Research, Ranchi, India, pp. 130-132.

Jolly, M.S., Chowdhury, S.N. and Sen, S.K. (1975), Non-mulberry sericulture in India. Central Silk Board, Bombay, pp. 1-89.

Jolly, M.S., Sen, S.K., Sonwalkar, T.S. and Prasad, G.K. (1979), Non-mulberry silks. FAO Agriculture Services Bulletin, Reprinted Central Silk Board, Bangalore, **4** : 1-178.

Joardar, O.I. and Islam, W. (1985), A short report on Bangladesh ecotype of tasar silkworm *A. mylitta*. Ind. J. Seric., **24** (1) : 36-39.

Joshi, K.L. (1987), Progression factor for growth in eri silkworm *Philosamia ricini* Hutt. (Lepidoptera : Saturniidae) in relation to diet. Ind. J. Seric., **24** (2) : 98-99.

Kapila, M.L. (1974), Studies on the emergence and voltinism of the spring crop cocoons of *A. proylei*. Proc. 1st Int. Sem. Non-mulberry silks, Ranchi, India, pp. 72-75.

Kameyama, T. and Kawasugi, S. (1965), Water soluble vitamin requirements of eri silkworm. Bull. Seric., Expt. Sta., Tokyo, **19** : 517-529.

Kanekatsu, R., Kitagawa, R. and Yanagisawa, Y. (1991), Mechanism of disaccharide utilization in the midgut of silkworm *Bombyx mori* using semisynthetic diets containing chromium trioxide as a marker. J. Seric. Sci., **60** (1) : 55-61.

Kar, R., Majumdar, S.K., Datta, R.N. and Sen, S.K. (1994), Mulberry cultivation in the acid soils of Bihar and West Bengal. Indian Silk, **33** (2) : 19-21.

Kato, H. (1990), Development of crimped silk yarn finished by salt shrinking and fibre dispersion treatment. J. Seric. Sci., **9** (4) : 271-279.

Kato, Y. and Sakate, S. (1981), Studies on summer diapause of *Antheraea yamamai* (Lepidoptera : Saturniidae). III. Influence of photoperiod of larval stage. Appl. Ent. Zool., **16** (4): 499-500.

Kato, Y., Nakamura, T. and Takeuchi, T. (1991), Haemaggultinating activity of haemolymph of *Bombyx mori* treated with terpenoid imidazole. J. Seric. Sci., **60**(3) : 208-213.

Kebschull, D. (1997), Silk Industry in Bangalore. German News, **38** (Jan-Feb): 17-18.

Kitamura, A., Chen, K. and Sibamata, A. (1990), Thermodynamic parameters in the diffusion process of organic solvents into silk. J. Seric. Sci., **59**(4) : 259-264.

Kogure, M. (1933), The influence of light and temperature on certain characters of the silkworm, *B. mori*. Dept. Agri. Kyshu Imp. Univ., **4** (6) : 1-93.

Kohli, R.K., Jolly, M.S. and Khan, A.N. (1969), Foliar constituents of the food plant of tasar silkworm *A. mylitta*. Indian Forester, **95** : 614-617.

Koidsumi, K. (1954), Negative correlation in the weight between the final larval exuviae and the cocoon silk in some saturniid insects. Jap. J. Appl. Zool., **18**(4) : 151-154.

Krishnamurthy, C.S. (1974), Tasar industry in Madhya Pradesh : breakthrough with block plantation. Souvenir, Central Silk Board, Bombay, XII (12) : 31-33.

Kumararaj, S. (1968), Further studies on double cocoons of *Bombyx mori* L. Ind. J. Seric., 7 (1) : 70-71.

Kumar, P.A. and Choudhury, S.K. (1981), Specific gravity of fertile eggs of *A. proylei*. Annu. Rept. Central Tasar Research, Ranchi, India, pp. 109.

Kumar, P.A. and Sinha, S.S. (1974), Effect of photoperiod and temperature on pupal diapause in *A. proylei*. Proc. Ist Int. Sem. Non-mulberry silks, pp. 135-142.

Ledoux, A. (1949), *Oecophylla*, nests, mating flights.Ann. Sci. Nat. Zool., **12** : 313-461.

Ledoux, A. (1958), Arboreal nests. Proc. Xth Int. Cong. Entomol., Montreal, **2**:521-528.

Lees, A.D. (1955), The physiology of diapause in arthropods. Cambridge Monogram. Cambridge Univ. Press, London. Exp. Biol., **4** : 1-151.

Lefroy, H.M. (1971), Indian insect life. Today and tomorrow Printers and Publishers, New Delhi, pp. 480.

Legay, J.M. (1958), Recent advances in silkworm nutrition. Ann. Rev. Ent., **3** : 75-86

Mansingh, A. and Smallman, B.N. (1967), Effect of photoperiod on the incidence of physiology of diapause in two Saturniids. J. Insect Physiol, **13** : 1147-1162.

Mishra, S.D. (1962), Nutritional ecology of the clearwinged grasshopper, *Camnula pellucida* (Scudder) (Orthoptera : Acrididae). Memoir Indian Museum (14): 87-174.

Misra, S.D. (1986), Ericulture. Lectures on sericulture, Bangalore, pp. 187-204.

Mohanty, P.K. (1991), Studies on some aspects of biology and ecology of *Antheraea paphia* Linn. (Lepidoptera : Saturniidae). Proc. 78th Ind. Sci. Cong., Indore, Pt.III, pp. 93.

Mohanty, P.K., Das, M.C., Mishra, C.S.K. and Nayak, B.K. ((1992), Seasonal Impact on life span of Tasar moth *Antheraea mylitta*. Proc. 79th Ind. Sci. Cong., Baroda, Pt. III, pp. 65.

Mohanty, P.K. and Nayak, B.K. (1993), Studies on starvation stress on hatchlings of tasar insect *Antheraea mylitta* Drury. Proc. 80th Ind. Sci. Cong., Goa, Pt. III, pp. 8.

Mohanty, P.K., Mishra, C.S.K., Nayak, B.K. and Das, M.C. (1993), Seasonal impact on the life span of the Indian tasar moth *Antheraea mylitta* Drury. Ad. Bios., **12**(II) : 81-84.

Mohanty, P.K. (1994), Sex ratio of coloured cocoons in different seasons of tasar silk insect *Antheraea paphia* Linn. (Lepidoptera : Saturniidae). Proc. 81st. Ind. Sci. Cong., Jaipur, Pt.III., pp. 44.

Mohanty, P.K. (1994), Tasar silk : the pride of Orissa. Orissa Review, **51** (5) : 18-20.

Mohanty, P.K. and Behera, M.K. (1995), Tasar cocoon and Tribal belief. Indian Silk, **34**(4) : 44.

Mohanty, P.K. (1995), Wild Tasarmoth and its conservation. Zoos' Print. **10**(10) : 1-3.

Mohanty, P.K. and Behera, M.K. (1995), Some observations on hammock formation in tasar silk insect *Antheraea paphia* Linn. (Lepidoptera : Saturniidae). Ad. Bios., **15**(1) : 1-4.

Mohanty, P.K. and Behera, M.K. (1996), Mortality of tasar silk worm *Antheraea paphia* Linn. (Lepidoptera : Saturniidae) due to pebrine (*Nosema* sp.) infection. Environment and Ecology, 14(2) : 358-360.

Mohanty, P.K. (1997), Species diversity in tasar silkmoth and its future scenario. Zoos' Print, **12**(2) : 5-6.

Mohanty, P.K. and Behera, M.K. (1997), Emergence and mortality behaviour of tropical wild tasarmoth *Antheraea paphia* Linn. Proc. 84th Ind. Sci. Cong., New Delhi, Pt.III. pp. 18.

Moroshi, S. (1957), Physiological studies on moltinism and voltinism in *Bombyx mori*, A new hormonal antagonistic balance theory on the growth. Jap. Soc. Promotion of Science, pp. 1-202.

Munirajappa, K. and Jayaramaiah, V. (1988), Morphological, ecological and meiotic studies in food plants of tropical tasar silkworm (*Antheraea mylitta* D.). Sericologia, **28**(2) : 273-280.

Muslim, M.M. (1990), Artificial hatching of silkworm eggs after chilling. Pak. J. For., **40**(2) : 167-172.

Nakazima, F. (1974), A short history of *A. yamamai*. Proc. Ist Int. Sem. Non-mulberry silks, Ranchi, India, pp. 48.

Nanavaty,M. (1965), Silk from grub to Glamour. Paramount Publishing House, Bombay, pp. 192-209.

Nayak, B.K. (1988), Standardisation of norms for grainage activities and rearing of tasar silk insect *Antheraea mylitta* Drury in different seasons at the trivoltine tasar areas. Annu. Rept. St. Seric. Res. Stn. Ori., **1** : 63-66.

Narasimhanna, M.N. and Jolly, M.S. (1969), Developmental morphology of tasar silkworm *A. mylitta* D. Ind. J. Seric., **8** : 1-3.

Narasimhanna, M.N., Ahsan, M.M. and Chakraborty, D. (1979), A field oriented technique of rearing chawki tasar silkworm of *A. mylitta* D. Ind. J. Seric., **18**(1) : 60-61.

Narasimhanna, M.N., Sen, S.K., Jha, M.K. and Vijaykumar, H.V. (1981), Performance of pure lines and hybrids of tasar silkworm *A. mylitta* Drury. Ind. J. Seric., **20**(1) : 1-6.

Narayanan, E.S. and Sonwalker, T.N. (1967), A simple laboratory method to find out the specific gravity of silk fibres. Ind. J. Seric., **6** : 94-98.

Narayan Prakash, R., Periasamy, K. and Radhakrishnan. S. (1984), Studies on the double cocoon formation in two races of *Bombyx mori* L. (Lepidoptera : Saturniidae). Sericologia, **24**(2) : 167-174.

Narayan Prakash, R., Periasamy, K. and Radhakrishnan, S. (1985), Effect of dietary water content on food utilization and silk production in *B. mori*. Ind. J. Seric., **24**(1) : 12-17.

Nayak, B.K. and Dash, M.C. (1991), Environmental regulation of voltinism in *Antheraea mylitta* Drury (Lepidoptera : Saturniidae), the Indian tasar silk insect. Sericologia, **31**(3) : 479-486.

Nayak, B.K., Pradhan, K.C. and Satpathy, B.N. (1985), Studies on Asan (*Terminalia tomentosa*) leaf consumption by successive instars and during the entire larval period of *Antheraea mylitta* Drury. Sericologia, **25** (1) : 26-31.

Nayak, B.K., Dash, A.K., Das, P.K., Sasmal, A.K. and Satpathy, B.N. (1986a), Sex association in doubled cocoons of tasar silk insect *Antheraea mylitta* Drury (Lepidoptera : Saturniidae). Sericologia, **26**(3) : 285-290.

Nayak, B.K., Gupta, M.L., Dash, A.K. and Guru, B.C. (1986b), Sex relation in chupi cocoon of tasar silk insect *Antheraea mylitta* Drury (Lepidoptera : Saturniidae). Science and Culture, **52** (10): 349-350.

Nayak, B.K., Gupta, M.L., Guru, B.C. and Satpathy, B.N. (1987), Towards classification of cocoons in tasar silk insect *Antheraea mylitta* Drury (Lepidoptera : Saturniidae). Sericologia, **27**(3) : 505-512.

Nayak, B.K., Dash, A.K., Guru, B.C. and Satpathy, B.N. (1988), Sex association in bishellate cocoons of tasar silkworm *Antheraea mylitta* Drury. Entomon, **13**(1): 91-93.

Negi, B.K. and Sengupta, A.K. (1994), Bark eater of Muga food plants. Indian Silk, **33**(2): 45-46.

Nigam, M.P., Das, P.K., Rana, B. and Goel, R.K. (1985), Preservation of seed. Annu. Rept., Central Tasar Research, Ranchi, India, pp. 377-382.

Nuriev, E.S.H., Kulikov, A.V. and Mamedov, S.H.V. (1989), Correlation between duration and intensity of silk ESR signal at thermoxidative destruction. Biofizika, **34**(4) : 561-563.

Ochiai, S. (1962a), Studies on the development of eri silkworm *Philosamia cynthia* Boisduval with special reference to the development of silk gland. J. Fac. Agric. Tamagawa Univ., Tokyo, **7** (1) : 2-29.

Ochiai, S. (1962b), Embryological studies particularly studies on the development of silk gland comparing with that of silkworm. J.Fac. Agric. Tamagawa Univ., Tokyo, 7(3) : 2-27.

Ojha, N.G. and Tanti, B. (1981), Studies on grainage performance of Sukinda and Sarihan. Annu. Rept., Central Tasar Research, Ranchi, India, pp. 223-226.

Ojha, N.G., Singh, M.K., Sharan, S.K., Sinha, B.R.R. Pd. and Sinha, S.S. (1996), Tropical Tasar Egg Transportation Basket. Indian Silk, **35** (2 and 3) : 27-28.

Opender, K., Tikku, K., Saxena, B.P. and Atal, O.K. (1979), Growth and silk production in *B. mori* L. fed on three varieties of mulberry. Ind. J. Seric., **18**(1) : 1-5.

Panda, N. (1965), Mating of tasar silk moth *A. paphia* in captivity. Proc. 52nd Ind. Sci. Cong., Pt. III, pp. 380.

Pant, R. and Agarwal, H.C. (1965), Changes observed in the free aminoacid content of *P. ricini* pupal haemolymph during metamorphosis. Ind. J. Expt. Biol., 3 : 133-136.

Pant, R. and Morris, I.D. (1972), Variation in Glycogen free sugars in fat body of *P. ricini* during development. J. Biochem., Tokyo, **71** : 1-8.

Patil, G.M. and Savanurmath, C.J. (1989), Can tropical silkworm *Antheraea paphia* Linn. be reared indoors? Entomon, **14** (3-4) : 217-226.

Parthasarthy, B.S. (1972), Studies on the foliar nutrition of *T. arjuna*, a primary food plant of tasar silkworm. Annu. Rept., Central Tasar Research, Ranchi, India, pp. 233-238.

Periasamy, K. (1986), Problems and prospects of sericulture. Lectures on sericulture, Bangalore, pp. 98-107.

Poonia, F.S. (1978), Studies on food utilization and rate of growth during the development stages of eri silkworm, *Philosamia ricini* Hutt. (Lepidoptera : Saturniidae). Ind. J. Seric., **17** : 48-60.

Quo - Fn and Hsiang. P. (1965), The role of the brain and prothoracic glands in the development and differentiation of adult characters and internal reproductive system in the Indian silkworm *P. ricini*. Acta. Zool. Sin., Peking, **16** : 175-186.

Raju, P.J. and Krishnamurthy, N.B. (1987), Evolving a multivoltine white with short larval duration in silkworm *B. mori* L. Sericologia, **27**(1) : 65-76.

Rajendra Singh, Y. and Prasad, B. (1987), Correlation among body weight, pupal weight and fecundity in eri silkworm *Philosamia ricini* Hutt. (Lepidoptera : Saturniidae). Sericologia, **27**(1) : 49-64.

Rao, G.S. (1976), Degeneration of mugaworms. Indian Silk, **15** : 11-13.

Rao, R.A. Nagaraja and Rao, L.S. Prahalada (1961), A note on the occurrence of double cocoons in silkworm *Bombyx mori* L.J. Silkworm, **13**(4) : 341-346.

Richards, O.W. and Davis, R.G. (1977), Imm's Text Book of Entomology (10th Ed.), Chapman and Hall, London, Vol. I, pp. 368.

Ronald, C. (1988), The world silk market : Today and tomorrow. Souvenir Ist Int. Cong. Tropical Sericultural Practices. 18-23 Feb., Bangalore, pp. 73-87.

Roychowdhury, B.N. (1974), Effect of different mating duration and polygamy on the hatching percentage of *Philosamia ricini* Hutt. Proc. Ist. Int. Sem. Non-mulberry silks, Ranchi, India, pp. 37.

Samson, M.S. and Phukan, S.N. (1988), Influence of coloured substratum on egg laying of *A. assama* Westwood. Sericologia, **28**(3) : 423.

Samson, M.V., Baig, M., Sharma, S.D., Balavenkatasubbaiah, M., Sasidharan, T.O. and Jolly, M.S. (1990), Survey on the relative incidence of silkworm diseases in Karnataka. Ind. J. Seric., **29** (2) : 248-254.

Saratchandra, B. and Joshi, K.L. (1988), A note on rearing of eri silkworm, *P.ricini* (Saturniidae : Lepidoptera) on different varieties of tapioca (*Marihot utillissima* Phol.). Sericologia, **28**(3) : 425-429.

Satpathy, S.S., Sasmal, A.K. and Nayak, B.K. (1988), Report on grainage operation of tasar silk insect *Antheraea mylitta* Drury during the year 1987. Annu. Rept. St. Seric. Res. Stn. Ori., **5** : 67-74.

Saxena, Y.W. and Bhide, J.B. (1969), Sex combination in double cocoons of *Antheraea mylitta* Drury. Ind. J. Seric., **8** (1) : 34-35.

Sen, S.K. and Jolly, M.S. (1967), Incidence of mortality of tasar silkworm *A. mylitta* D. due to disease in relation to meteorological conditions and larval instars. Ind. J. Seric., 1 : 67-72.

Sen, S.K., Jolly, M.S. and Jammy, T.R. (1969), Disease of tasar silkworm *A. mylitta* D. (Lepidoptera : Saturniidae). Ind. J. Seric., **8** : 11-14.

Sen, S.K., Jolly, M.S. and Jammy, T.R. (1970), A mycosis in the Indian tasar silkworm *A. mylitta* D. caused by *P. citrunum* Thom. J. Invert. Pathol., **15** : 1-5.

Sen, S.K., Sengupta, A.K., Das, M.G. and Jolly, M.S. (1976), Statistics on genetic variability correlations, path coefficient analysis and discriminant fuction in *A. mylitta*. Ind. J. Seric., **15** 9-14.

Sengupta, K. (1986), Tasar silk industry in India. Central Tasar Research and Training Institute, Central Silk Board, Ranchi, India, pp. 1-27.

Sengupta, K., Prasad, D.N., Khare, R. and Ramkumar (1984), Effect of pruning on the vegetative growth and yield of leaves of primary tasar food plants Annu. Rept., Central Tasar Research, Ranchi, India, pp. 1-7.

Sengupta, K., Sumbli, S.N., Saddurin and Singh, M.K. (1986), High altitude cocoon preservation and general grainage operation. Annu. Rept., Central Tasar Research, Ranchi, India, pp. 396-399.

Shivkumar, G.R., Himanthraj, M.T., Rajan, R.K., Magadum, S.B. and Datta, R.K. (1994), Effect of transportation of bivoltine seed and cocoons on melting and egg production. Indian Silk, **33** (2) : 23-24.

Silk in India, Statistical Biennial (1988), (Edt. J. Sampath) Cental Silk Board, Ministry of Textiles - Govt. of India, Bangalore, pp. 1-144.

Silk in India, Statistical Biennial (1992), (Edt. J. Sampath) Central Silk Board, Ministry of Textiles - Govt. of India, Bangalore, pp. 1-159.

Silkman's Companion (1989), (Edt. J. Sampath) Central Silk Board, Ministry of Textiles - Govt. of India, Bangalore, pp. 1-107.

Silkman's Companion (1992), (Edt. J. Sampath) Central Silk Board, Ministry of Textiles - Govt. of India, Bangalore, pp. 1-78.

Singh, Y.N. (1972), Cephalothoracic nerves of the larva of castor silk moth *P. ricini* Hutt. (Lepidoptera : Saturniidae). Int. J. Insect Morph. Embryol., **2** : 101.

Singh, K.N. and Jolly, M.S. (1969), Air layering in *T. arjuna* W. & A. Indian Forester, **95** : 539-541.

Singh, S.S., Jolly, M.S., Singh, B. and Razdan, J.L. (1969), Studies on the effect of density population in tasar silkworm rearing (*A. mylitta* D.) Ind. J. Seric., **8** : 15-24.

Singh, G.P., Mathur, V.B., Kamble, C.K. and Dutta, R.K. (1994), Cold storage of bivoltine female silkworm moths. Indian Silk, **33** (2) : 25-26.

Sinha, B.N. (1981), Geography of Orissa. National Book Trust of India, New Delhi. pp. 1-58.

Sinha, B.R.R.P. and Jayaswal, J. (1972), Emergence, coupling, egg laying and hatching behaviour of *A. pernyi* and *A. roylei*. Annu. Rept., Central Tasar Research, Ranchi, India, pp. 258-260.

Sinha, A.K. and Jolly, M.S. (1971), Foliar constituents of the food plants of tasar silkworm *A. mylitta*. Indian Forester, **97** : 261-263.

Sinha, A.K., Sinha, U.S.P., Sinha, S.S. and Sengupta, K. (1990), Changes in free aminoacids in the hemolymph of pebrine infected larvae and pupae of *Antheraea mylitta* D. Ind. J. Seric., **29**(2) : 233-239.

Sonwalker, T.N. (1986), Silk reeling technology in India. Lectures on sericulture, Bangalore, pp. 205-219.

Sonwalker, T.N. and Jolly, M.S. (1985), Investigations on the influence of colour and size of Daba cocoons on reeling performance. Ind. J. Seric., **24** (1) : 1-6.

Sosagomez, D.R., Alves, S.B. and Marchini, L.C. (1991), Variation in the susceptibility of *Bombyx mori* to nuclear polyhydrosis virus when reared on different mulberry genotypes. J. Appl. Entomol., III (3) : 318-320.

Srivastava, A.K. and Ansari, M.O.A. (1979), Demonstration of new technique of cooking and reeling of tasar cocoons. Annu. Rept., Central Tasar Research, Ranchi, India, pp. 84.

Srivastava, A.D. and Misra, S.D. (1981), Effect of larval density on fecundity and productivity of erimoth *Philosamia ricini*. Geobios, **9** : 58-62.

Srivastava, A.D. and Misra, S.D. (1985), Effect of crowding on fecundity of erimoth *Philosamia ricini* Hutt. Ind. J. Seric., **15** : 41-43.

Srivastava, A.D., Misra, S.D. and Poonia, F.S. (1982), Effect of food deprivation on larval duration, cocoon (Shell) weight and fecundity of erimoth *Philosamia ricini* (Lepidoptera : Saturniidae). Ind. J. Seric., **22** : 11-15.

Subba Rao, G. (1978), New trends in eri and muga sericulture. Proc. Int. Silk Association, Bombay, pp. 43-46.

Subramanyam, K. and Narasimha Murthy, C.V. (1987), Effect of rest on the efficiency of male moth in relation to fecundity and hatching behaviour of silkworm eggs. Sericologia, **27** (4) : 677-684.

Sukumar, J. and Ramalingam, A. (1987), Epidemiology of sercospora leaf spot disease of mulberry. II. Leaf age and disease susceptibility. Sericologia, **27** (4) : 685-692.

Sullia, S.B. and Padma, S.D. (1987), Acceptance of mildew affected mulberry leaves by silkworm *B. mori* L. and its effects on cocoon characteristics. Sericologia, **27** (4) : 693-700.

Talukdar, J. N. (1961), Further report on the pebrine germ of *Philosamia ricini* Bsd. and its infective capacity on *B. mori* Linn. Curr. Sci., **30** : 28-29.

Talukdar, J.N. (1962) The dispersal of spores of a *Nosema* attacking *Philosamia ricini* Hutt. J. Insect. Pathol., **4** : 128-129.

Tanaka, Y. (1938), The feature of morphological tasar silkworm *A. mylitta* in Saigon. J. Seric. Sci., **24** : 24-118.

Tanaka, Y. (1950a), Studies on hibernation with special reference to photoperiod and breeding of Chinese tasar silkworm. J. Seric. Sci. Jap., **19** (4) : 358.

Tanaka, Y. (1950b), Studies on hibernation with special reference to photoperiodicity and breeding of the Chinese tasar silkworm. J. Seric. Sci. Jap., **19** (4) : 429.

Tanaka, Y. (1950c), Studies on hibernation with special reference to photoperiodicity and breeding of the Chinese tasar silkworm. J. Seric. Sci. Jap., **19** (4) : 520.

Tanaka, Y. (1951a), Studies on hibernation with special reference to photoperiodicity and breeding of the Chinese tasar silkworm. J. Seric. Sci. Jap., **20** : 191-201.

Tanaka, Y. (1951b), Studies on hibernation with special reference to photoperiodicity and breeding of the Chinese tasar silkworm. J. Seric. Sci. Jap., **20** : 132.

Tanaka, Y. (1956), Photoperiodic effect of alternated dark and light phases of diapause of *Antheraea pernyi*. Annu. Rept. Natl. Inst. Genetics, **6** : 35-36.

Tanaka, Y. (1956), Studies on polyhedrosis of *A. yamamai* Guer. and *A. pernyi* Guer. Bull. Nagnano Kon. Seric. Expt. Sta. Naganocity, Japan, pp. 68.

Tayade, D.S. (1982), Sex association in the double cocoons of *Bombyx mori* L. Ind. J. Seric., **21-22** : 70.

Tayade, D.S., Jamale, M.D. and Unchegaonkar, P.K. (1988), Effect of antibiotics on the growth of silkworm *Bombyx mori* L. Ind. J. Seric., **27**(2) : 69-72.

Tazima, Y. (1958), Report on sericultural industry in India, Central Silk Board, Bombay, pp. 1-50.

Thangavelu, K. (1986a), Muga silk production. Lectures on sericulture, Bangalore, pp. 166-174.

Thangavelu, K. (1986b), Golden muga silk. Lectures on sericulture, Bangalore, pp. 129-130.

Umeya, Y. (1939), A consideration on the wintering of the insects from a stand point on the life cycle of the silkworm. Zool. Mag., **51** (7) : 58-162.

Veeraiah, K. (1986), A criticial look at the mulberry cultivation in Karnataka. Lectures on sericulture, Bangalore, pp. 4-8.

Venugopala Pilai, S. and Krishnaswai, S. (1987), Adaptability of silkworm *B. mori* to tropical conditions. Studies on the effect of high temperature during later developmental stages of silkworm. Ind. J. Seric., **25** (1) : 63-71.

Visweswara Gowda, B.L., Narayanaswamy, T.K. and Devaiah, M.C. (1987), Effect of refrigeration on blue stage eggs and the freshly hatched larvae on survival and cocoon parameters in mulberry silkworm *B. mori* L. Sericologia, **27** (4) : 715-726.

Watanabe, K. (1924), Studies on the voltinism in the silkworm *B. mori.* Bull. Seric. Imp. Exp. Sta., **3** (8) : 46.

Watanabe, H. (1967), Development of resistance in the silkworm *B. mori* to peronal infection of a cytoplasmic polyhedrosis. J. Invert. Pathol., **9** : 474-479.

Wheeler, W.M. (1915), Cocoons. Ann. Ent. Soc. Amer., **8** : 332-342.

Williams, C.M., Adkisson, P.L. and Walcott, H.C. (1985), Physiology of insect diapause. The transmission of photoperiod signals to the brain of oak silkworm *Antheraea pernyi*. Biol. Bull., **128** : 497-507.

XIE, W. (1989), The ultrastructure of *Nosema* sp., a microsporidian parasite from *Philosamia cynthia*. Acta Zool. Sin., **35**(4) : 345-347.

Yamazaki, H., Nishimura, K. and Yamada. T. (1954), Studies on the influence of temperature upon the development of wild silkmoth. J. Soc. Agric. Met., Tokyo, pp. 23-25.

Yokoyama, T. (1962), Synthesized science of sericulture. Central Silk Board, Bombay, pp. 337-339.

Index

D

Standard Raw Silk Packages and Measures

India

1. Filiature raw silk

Book	:	2 Kg
Bale	:	20 Kg

2..Spun silk yarn

Bundle	:	2.5 Kg
Bale	:	50 Kg

3. Noil yarn

Bundle	:	5 Kg
Bale	:	45 Kg

Other Countries

Japanese/ Korean bale	:	1 picul -60 Kg or 132. 27 lb
Brazilian cartoon	:	30 Kg flat skeins
Weight of skeins standard for 20-22 denier raw silk	:	70 gms (approx.)

Units in Tasar Industry

1. Kahan - 1600 cocoons (Orissa)
2. Kahan - 1280 cocoons (Bihar)
3. Kahan - 1000 cocoons (M.P.)
4. Kahan - 4000 cocoon (Maharashtra)
5. 4 cocoons = 1 ganda
6. 20 ganda = 1 pon
7. 16 pons = 1 kahan (Bihar)
8. 20 pons = 1 kahan (Orissa)

Plates

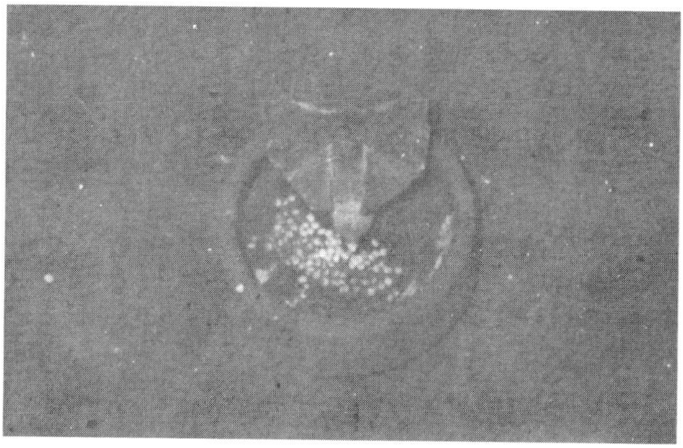

Plate 1 Eggs laid by female moth

Plate 2 First instar larvae hatched from eggs

Plate 3 Second instar larvae

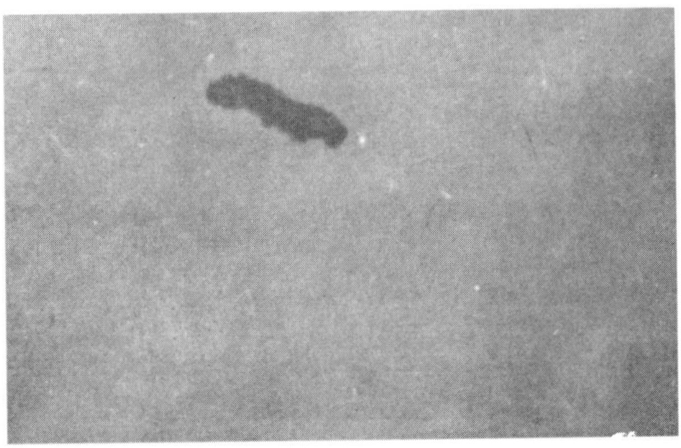

Plate 4 Third instar larva

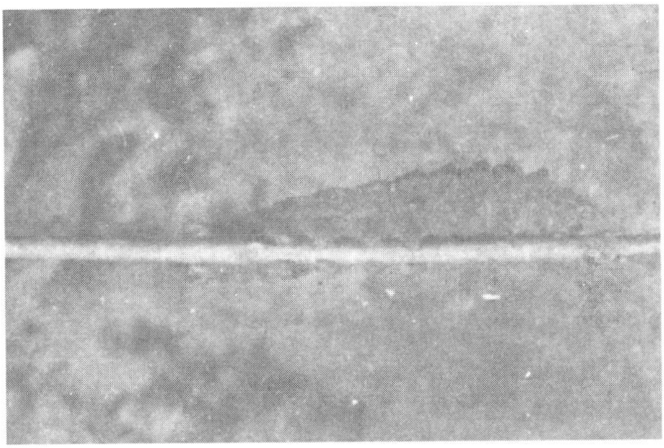

Plate 5 Fourth instar larva

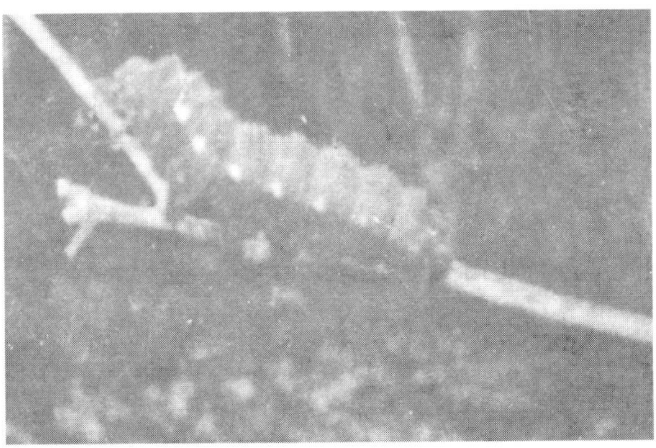

Plate 6 Fifth instar Larva

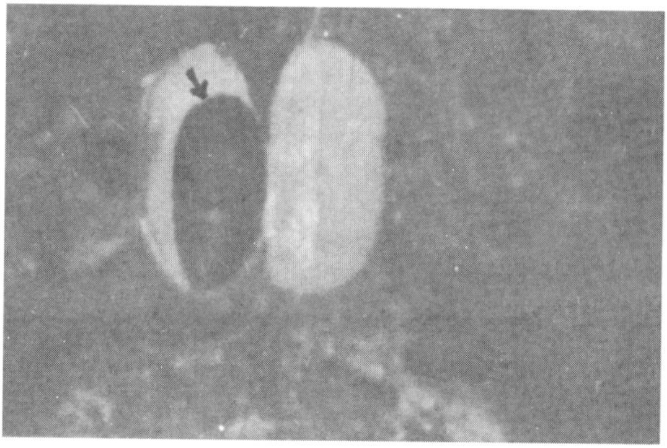

Plate 7 Pupa inside cocoon (Cocoon cut open)

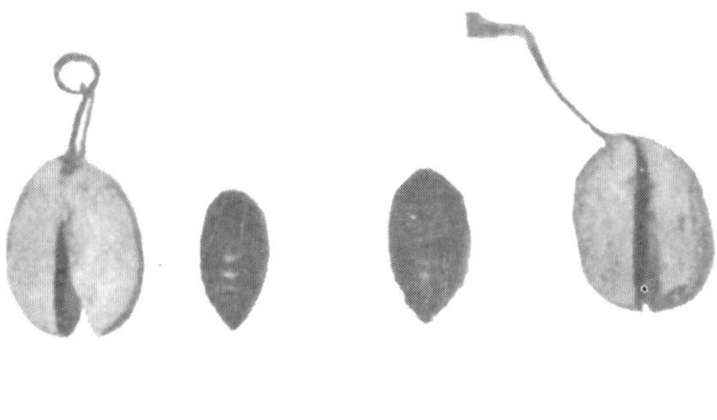

Plate 8 Male and female pupa from cocoon

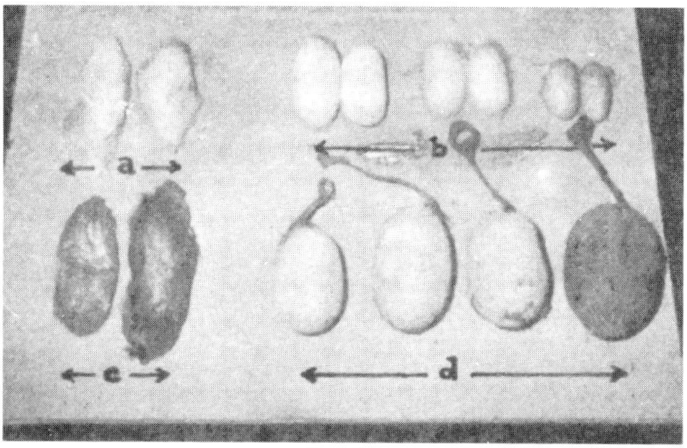

Plate 9 (a) Eri (b) Mulberry (c) Muga (d) Tasar cocoons

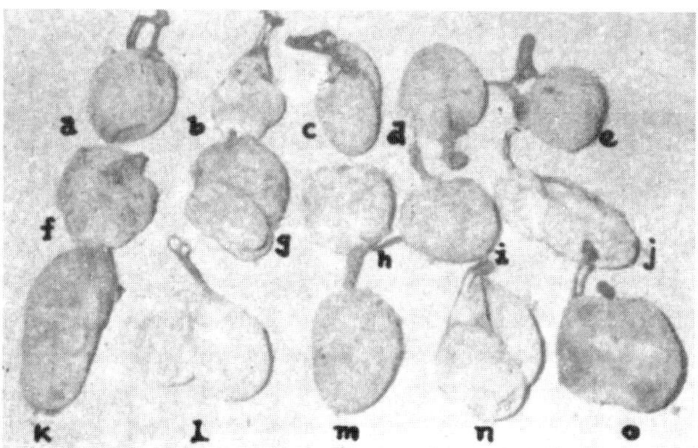

Plate 10 Abnormal cocoons

Plate 11 Multiloopate cocoons

Plate 12 Synloopate cocoons

Plate 13 (a) Male (b) Female tasar moth

Plate 14 Selection of seed cocoon

Plate 15 Cocoons garlanded in grainage

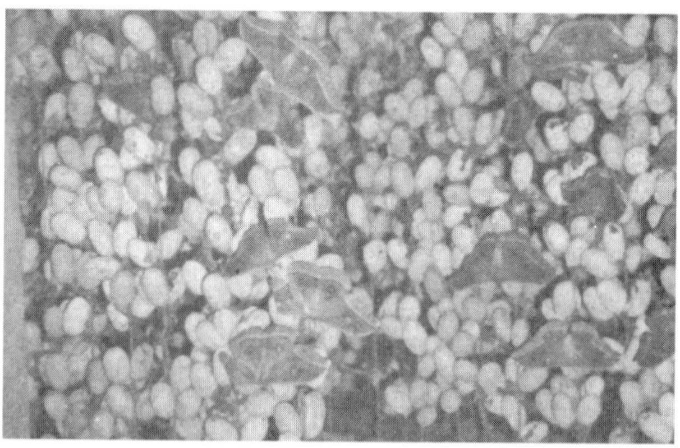

Plate 16 Emergence of moth from cocoon and pairing

Plate 17 Coupling between male and female

Plate 18 Captive coupling in grainage

Plate 19 Cleaning and sterilising and earthen pots

Plate 20 Earthen pots collected after egg-laying

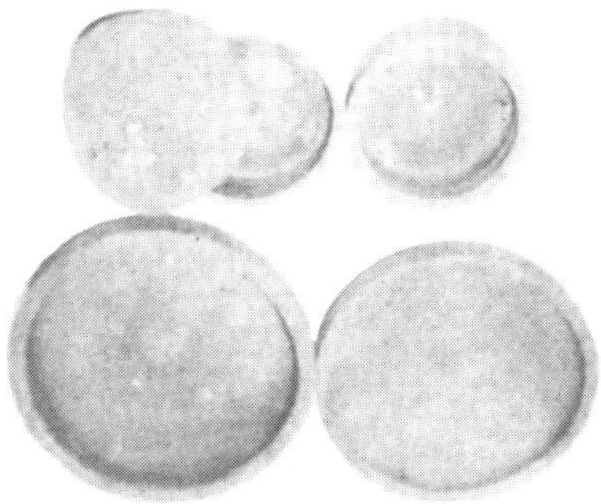

Plate 21 Earthen pots for tasar egg-laying

Plate 22 Perforated plastic boxes for egg collection

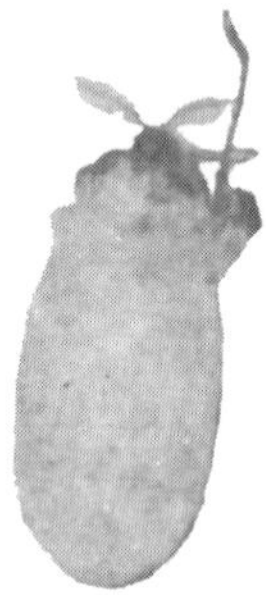

Plate 23 Partial emergence of tasar moth

Plate 24 Complete emergence from cocoon

Plate 25 Exuviae inside cocoon after emergence

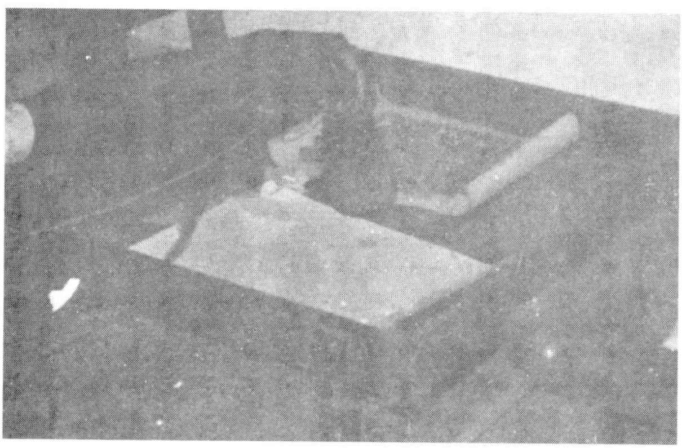

Plate 26 Washing of tasar eggs

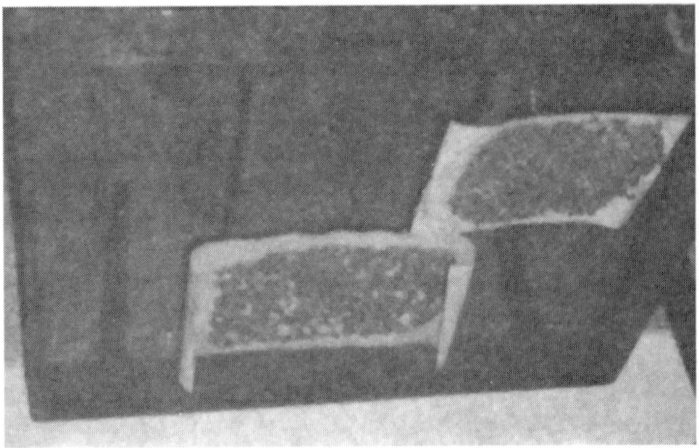

Plate 27 Drying of eggs on blotting paper

Plate 28 Disease free layings in cloth packet